特别鸣谢

四川省科学技术厅
对本书出版提供的支持

Special thanks to
the Science and Technology Department
of Sichuan Provincial
for supporting the publication of this book

成都大熊猫繁育研究基地
CHENGDU RESEARCH BASE OF GIANT PANDA BREEDING

成都大熊猫繁育研究基地　版权所有
未经版权所有方许可，不得对本书内容进行转载、翻印，
或以其他手段进行传播。

Copyright © 2024
Chengdu Research Base of Giant Panda Breeding

All rights reserved. No part of this publication maybe
reproduced, stored in a retrieval system, or transmitted
in any form or by any means, electronic, photocopying,
recording, or otherwise, without the prior written
permission of the copyright owner.

熊猫世界
The World of Panda

金 双 编著
JIN SHUANG

四川少年儿童出版社

图书在版编目（CIP）数据

熊猫世界：汉英对照 / 金双编著 . — 成都：四川少年儿童出版社，2024.4（2024.9 重印）
ISBN 978-7-5728-0840-1

Ⅰ. ①熊… Ⅱ. ①金… Ⅲ. ①大熊猫－普及读物－汉、英 Ⅳ. ① Q959.838-49

中国国家版本馆 CIP 数据核字 (2024) 第 061760 号

熊猫世界
The World of Panda
金　双　编著

出 版 人	余　兰
责任编辑	李明颖　蒲幼鲲
美术编辑	苟雪梅
装帧设计	薛先良（良友设计）
插画绘制	杨　彬　郎佳琪
责任校对	王默志
责任印制	袁学团
出　　版	四川少年儿童出版社
地　　址	成都市锦江区三色路 238 号
电　　话	028-86361727（发行部）
网　　址	http://www.sccph.com.cn
网　　店	http://scsnetcbs.tmall.com
经　　销	新华书店
印　　刷	深圳市深教精雅印刷有限公司
成品尺寸	300mm×250mm
开　　本	12
印　　张	$21\frac{1}{3}$
版　　次	2024 年 6 月第 1 版
印　　次	2024 年 9 月第 2 次印刷
书　　号	ISBN 978-7-5728-0840-1
定　　价	298.00 元

版权所有　翻印必究
若发现印装质量问题，请及时与市场发行部联系调换
地　　址：成都市锦江区三色路238号新华之星A座23层四川少年儿童出版社市场发行部
邮　　编：610023

科学顾问
Scientific Advisors

侯蓉、兰景超、黄祥明、吴孔菊、李明喜、刘玉良、
齐敦武、罗娌、沈富军、聂溟飞、杨奎兴、许萍、
魏玲、李洁、张玉均、杨文光、巫嘉伟、罗瑜

Hou Rong, Lan Jingchao, Huang Xiangming, Wu Kongju, Li Mingxi,
Liu Yuliang, Qi Dunwu, Luo LI, Shen Fujun, Nie Mingfei, Yang Kuixing, Xu Ping,
Wei Ling, Li Jie, Zhang Yujun, Yang Wenguang, Wu Jiawei, Luo Yu

特约审稿
Special Reviewers

吴樱、唐亚飞、向波、徐星煜、王思敏、郭晶、王益梅、
刘菲、陈丽、李雯婧、罗珍、孙杉、肖爽、魏嘉、郭俊良、
杨林艳、李雨曦、罗琴、谭琴、唐薇、杨欣然、谭棋、张庚、
赵昕、张琰杰、许琴、尹红、李丹、王英、黄嘉雯

Wu Ying, Tang Yafei, Xiang Bo, Xu Xingyu, Wang Simin, Guo Jing, Wang Yimei,
Liu Fei, Chen LI, Li Wenjing, Luo Zhen, Sun Shan, Xiao Shuang, Wei Jia, Guo Junliang,
Yang Linyan, Li Yuxi, Luo Qin, Tan Qin, Tang Wei, Yang Xinran, Tan Qi, Zhang Geng,
Zhao Xin, Zhang Yanjie, Xu Qin, Yin Hong, Li Dan, Wang Ying, Huang Jiawen

文学指导
Literary Supervisor

秦丽

Qin Li

英文审定
English Revision

吴樱

Wu Ying

图片提供
Picture Providers

成都大熊猫繁育研究基地
Chengdu Research Base of Giant Panda Breeding

西华师范大学
China West Normal University

中国大熊猫保护研究中心
China Conservation and Research Center for the Giant Panda

海峡（福州）大熊猫研究交流中心
Strait (Fuzhou) Giant Panda Research and Exchange Center

崔凯、雍严格、张玉均
Cui Kai, Yong Yan'ge, Zhang Yujun

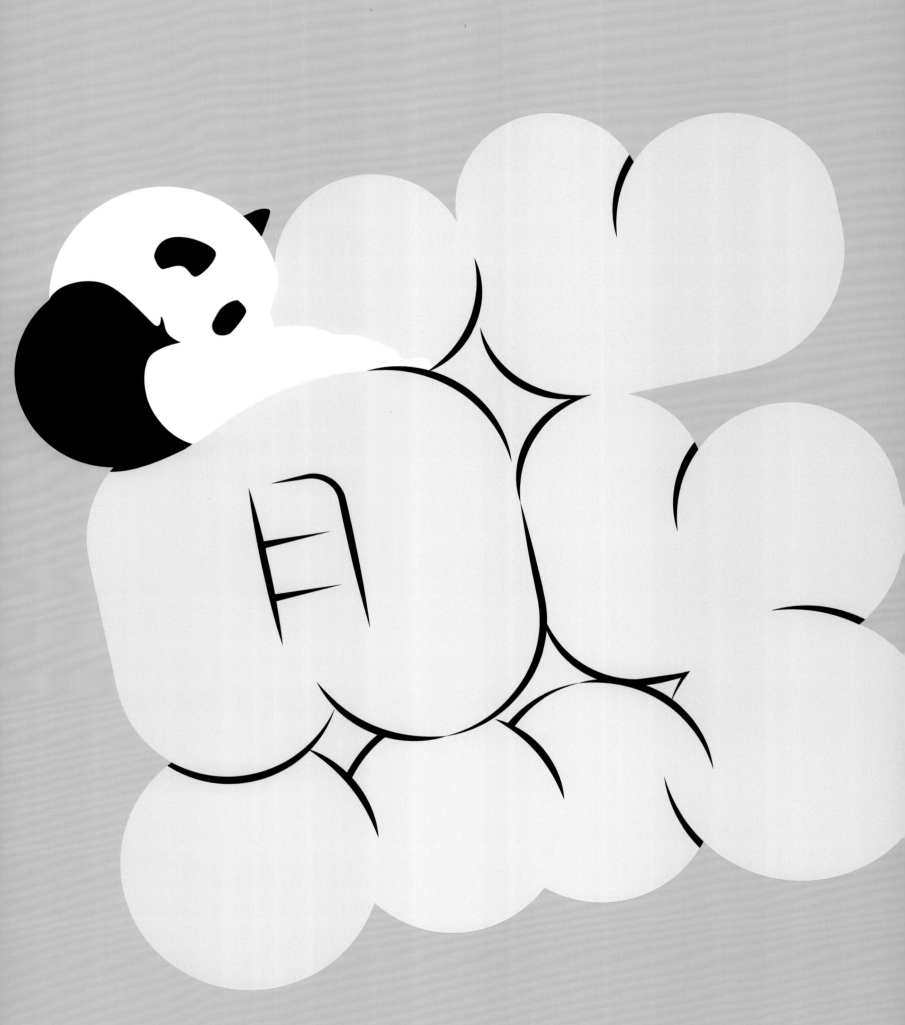

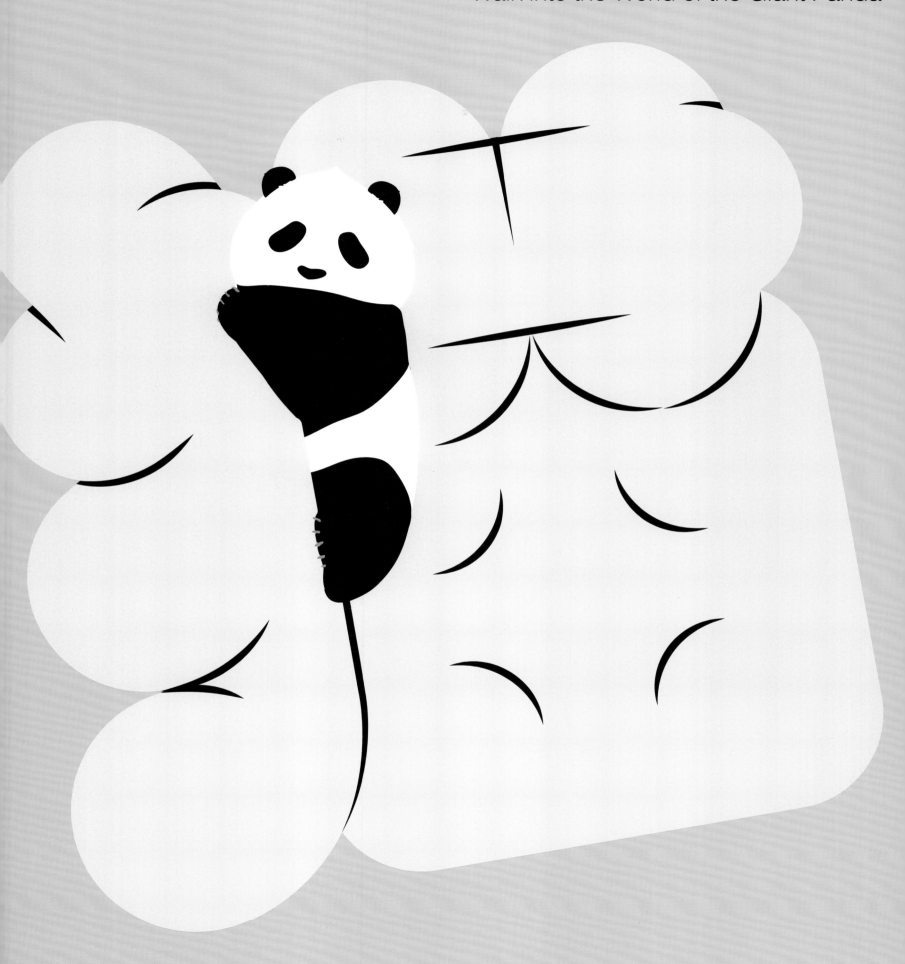

走进大熊猫的世界
Walk into the World of the Giant Panda

探索大熊猫的奥秘
Explore the Mysteries of the Giant Panda

我是"国宝"大熊猫

我有圆圆的脸颊,大大的黑眼圈,
胖嘟嘟的身体,黑白分明的皮毛,
标志性的内八字行走方式,还有锋利的爪子,
被人类称为世界上最可爱的动物之一。
我们家族已在地球上生存了至少 800 万年,
是国家一级重点保护野生动物,
世界自然基金会形象大使,
也是世界生物多样性保护的旗舰物种。

I am a giant panda, China's national treasure.

I'm chubby with distinctive black and white fur, round cheeks, big dark circles around my eyes, and sharp claws.
Coupled with my iconic pigeon-toed way of walking, my features have made me one of the most adored animals around the globe.
Surviving on Earth for at least 8 million years, we are a key wildlife species under China's first-class protection, the iconic logo of the World Wide Fund for Nature (WWF), and a flagship species for the conservation of the world's biodiversity.

大熊猫

别　　名：貔貅、貘、驺虞、猫熊、黑白熊、花熊、竹熊、食铁兽
学　　名：*Ailuropoda melanoleuca*
英文名：Giant Panda
出生地：**中国**

分类地位　　界：动物界
　　　　　　　门：脊索动物门
　　　　　　　纲：哺乳动物纲
　　　　　　　目：食肉目
　　　　　　　科：熊科
　　　　　　　属：大熊猫属
　　　　　　　种：大熊猫

世界自然保护联盟濒危物种红色名录（IUCN）：**易危（VU）**
国家重点保护野生动物名录：**中国国家一级重点保护野生动物**

Giant Panda

Aliases: Pixiu, Mo, Zouyu, Bear Cat, Black-white Bear, Mottled Bear, Bamboo Bear, Iron-eater
Scientific name: *Ailuropoda melanoleuca*
English name: Giant Panda
Place of birth: **China**

Taxonomic Status
Kingdom: Animalia
Phylum: Chordata
Class: Mammalia
Order: Carnivora
Family: Ursidae
Genus: *Ailuropoda*
Species: *Ailuropoda melanoleuca*

IUCN Red List of Threatened Species: **Vulnerable (VU)**
List of National Key Protected Wild Animals:
National First-class Protected Wild Animal

序言
>> Introduction

大熊猫曾一度濒临灭绝，20世纪80年代全球野生大熊猫仅存1100余只。多年来，社会各界在大熊猫种群及其栖息地的保护上投入了大量资源，做了全面细致的工作，取得了举世瞩目的成就，我国大熊猫栖息地受保护面积从139万公顷增长至258万公顷，大熊猫野外种群数量从20世纪80年代的1100余只增至近1900只，实现了大熊猫栖息地保护面积和野外种群数量"双增长"。世界自然保护联盟将大熊猫的受威胁等级由"濒危"调整为"易危"。成绩令人振奋，但并不意味着大熊猫保护工作从此可以高枕无忧。第四次大熊猫野生资源调查的结果显示，野生大熊猫现今仅存于青藏高原东缘的6大山系中，形成了30余个相互隔离的种群，其中约20个相对孤立的小种群存在较高的灭绝风险。野生大熊猫的生存仍然面临巨大挑战，大熊猫保护事业还有很长的路要走。

"未来属于青年，希望寄予青年"。普及生态科学知识，引导青少年树立生态文明理念，对于濒危野生动物保护事业的发展至关重要。

科普创作是科学普及工作的重要基础。成都大熊猫繁育研究基地丰硕的科研繁育成果，为科普创作提供了丰富且珍贵的素材。本

Giant pandas were once on the verge of extinction. In the 1980's, just more than 1,100 giant pandas existed in the wild across the globe. Over the years, thanks to tremendous investments in protecting the species and their habitats and to comprehensive and meticulous work done by all sectors of society, remarkable achievements have been made, with the protected area of giant panda habitat growing from 1.39 million hectares to 2.58 million hectares, the number of wild populations jumping from just over 1100 in the 1980s to nearly 1,900, achieving a "double growth" of the protected area and wild population of giant pandas. The International Union for Conservation of Nature (IUCN) has downgraded giant pandas from "endangered" to "vulnerable". The encouraging results do not necessarily mean the conservation efforts should be scaled back. The results of the Fourth Survey on Wild Giant Pandas show that wild giant pandas now only inhabit 6 mountain systems on the eastern edge of the Qinghai-Tibet Plateau, which has given rise to more than 30 fragmented populations. Among them, about 20 relatively small populations face a higher risk of extinction, which means extinction-level challenges are still lingering; there is still a long way to go to conserve giant pandas.

"As the future belongs to the youth, hope rests with them." Popularizing scientific knowledge and raising young people's awareness of ecological civilization is crucial to facilitating endangered-wildlife conservation.

Innovative ways of promoting conservation

education underpins the process. The fruitful results of breeding research by the Chengdu Research Base of Giant Panda Breeding provide abundant and precious resources for conservation education. For example, some of the contents in this book, such as "Discerning Strategies for Eating Bamboo", "Breeding and Growth of the Giant Panda", "More Scientific Research Breakthroughs" and so on, are the brainchild of the Base; others like "Er Yatou went through a difficult birth", "Mao Dou without pants", and "A keeper's day" are the personal experience of frontline staff. It can be said that this book is not a collection of existing material, but a unique creation based on first-hand information. In addition, author Jin Shuang's lively style of writing combined with ample pictures, original illustrations, and accompanying text, provides readers with a pleasant experience. In a nutshell, this book is a "museum of giant pandas on paper" that integrates science, knowledge, and fun into one.

Although ecological civilization construction is in full swing, giant panda conservation remains a large challenge. I hope that this book will play a positive role in spreading scientific knowledge of giant pandas and raising people's awareness of wildlife conservation.

Yin Zhidong
Director of the Chengdu Research Base of Giant Panda Breeding

书中的一些内容，比如"食竹攻略""孕育和成长""科研突破、再突破"等，就是熊猫基地自己的科研成果；"'二丫头'生宝宝难产了""'毛豆'脱裤子了""饲养员的一天"等，则是一线工作人员的亲身经历。因此，本书不是搜集汇总，而是基于一手资料的独家创作。此外，作者金双文笔生动活泼，书中丰富的资料图片、原创插画和文字相互呼应，为读者提供了一种轻松愉快的阅读体验。总的来说，本书堪称一座融科学性、知识性、趣味性于一体的"纸上大熊猫博物馆"。

生态文明建设方兴未艾，大熊猫保护事业任重道远。愿本书能够为普及大熊猫科学知识、传播野生动植物保护理念起到积极作用。

成都大熊猫繁育研究基地主任

目录
>> Content

前言	Foreword	016

第一章　熊猫前传　Chapter 1　Tracing Origins — 022
大熊猫的演化史　Evolution History of the Giant Panda — 024

第二章　竹林隐士　Chapter 2　Among the Pandas — 046
大熊猫的身体　The Giant Panda Body — 048
大熊猫的感官和"语言"　Senses and Language of the Giant Panda — 062
大熊猫的吃与憩　Diet and Rest of the Giant Panda — 066
大熊猫的孕育和成长　Breeding and Growth of the Giant Panda — 078
大熊猫的生活习性　Giant Panda Habits — 088
大熊猫的科属　Giant Panda Family and Genus — 090

第三章　发现熊猫　Chapter 3　Discovery Trail — 102
古籍里的记载　Ancient Records — 104
科学发现大熊猫　Science-based Discovery of the Giant Panda — 108
大熊猫走向世界　Giant Panda Going Global — 114
大熊猫保护研究　Giant Pandas' Conservation and Research — 120
友谊的使者　Messenger of Friendship — 126
大熊猫国际明星　Giant Panda International Stars — 134

第四章	濒危年代　Chapter 4	Human Impact	140
	大熊猫的野外家园	Wild Home for Giant Pandas	142
	大熊猫栖息地选择的要素	Factors Concerning Giant Pandas' Habitat Selection	146
	大熊猫野生资源调查	Survey on Wild Giant Pandas	156
	大熊猫的生存危机	Survival Crisis Facing Giant Pandas	164
第五章	保护之路　Chapter 5	Helping Hands	168
	大熊猫保护的法制底气	Laws Underpinning Giant Panda Conservation	170
	两种保护方式	Two Methods of Conservation	172
	专业又用心的"熊猫人"	The Professional and Attentive Panda Conservationists	182
	大熊猫的疾病与救治	Giant Panda Diseases and Treatments	190
	大熊猫野化放归	Giant Panda Rewilding	196
	从濒危到易危	Downgraded from "endangered" to "vulnerable"	202
第六章	生态家园　Chapter 6	Taking Actions	204
	中国的生物多样性	Biodiversity in China	206
	保护大熊猫的重要意义	The Importance of Conserving Giant Pandas	210
	大熊猫的动植物邻居	Giant Panda's Flora and Fauna Neighbors	214
第七章	创享未来　Chapter 7	Join the Future	232
	大熊猫的影响力	Influence of Giant Pandas	234
	保护环境就是保护大熊猫	Protecting the environment is to conserve Giant Pandas	244
后记及致谢		Postscript and Acknowledgments	253

前言 >> Foreword

不过瘾，不过瘾，真不过瘾！为了看一眼萌宝大熊猫，你从五湖四海赶来，在人山人海中排队 2 小时，看"猫" 3 分钟……

想过瘾？安排，那必须安排！想不想摸摸熊猫皮毛是什么手感？要不要通过互动装置变身"国宝医生"，上手给大熊猫照个"X 光"，看看大熊猫的骨骼、肌肉和其他身体部位到底是啥样？或者，来考古沙坑里挖呀挖呀挖，挖出仿真骨头和牙齿，再通过旁边的"扫描仪"鉴定一下它们是不是出自你最爱的大熊猫？来吧，这里能让你的愿望通通得到满足。

坐标：全球大熊猫"含量"最高的城市——成都，著名的大熊猫科研保护单位——成都大熊猫繁育研究基地，全球首座以大熊猫为主题的互动体验专题博物馆——成都大熊猫博物馆。

这是一座主打互动体验的现代博物馆，多种多样的视听设备和新技术手段被用心融入陈列设计中。在这里，你不仅能看、能摸、能听，还能闻（对，能闻！），各种互动装置能让你玩得不亦乐乎，沉浸式的游览体验会让你感觉进入了神奇而生动的熊猫世界。

逛大熊猫博物馆，最懂行的肯定是大熊猫本猫呀。那就请出我们的形象代言人——小川，一只有着红红脸蛋儿、头上顶着一簇毛的卡通大熊猫。小川的原型是因"抱大腿"视频走红、萌翻全球大熊猫粉丝的大熊猫"奇一"。"奇一"的生日是 2016 年 7 月 1 日，它的名字就来自生日的谐音。头顶长了一簇像 Wi-Fi 天线一样毛发

It's a kick to see giant pandas frolicking and lazily sunbathing! Some tourists travel from all over the world and wait for two hours to see the pandas.

We've listened to our fans and have exciting new changes. Would you like to be a panda keeper to feel the panda's fur? Would you like to become a panda doctor through an interactive installation and take an x-ray to see what their bones, muscles and other body parts really look like? Or, become an archaeologist and uncover simulated bones and teeth, and then scan them to identify whether they are from your favorite pandas? Come on, this place will satisfy all your desires to learn about pandas.

Location: The Chengdu Giant Panda Museum, the world's first panda-themed interactive experience museum at the Chengdu Research Base of Giant Panda Breeding, the famous scientific research and conservation institution, is located in Chengdu.

This is a modern museum featuring interactive experiences with a variety of audio-visual equipment and new technological forms of activities well incorporated into the display designs. Tourists can immerse themselves in various interactive devices, which enable them not only to see, touch, and hear, but also smell (yes, smell!) in the museum, as if they have entered the magical and vivid world of pandas.

No one knows the ropes of visiting the Giant Panda Museum better than giant pandas themselves. You'll meet our brand ambassador – Xiao Chuan, a cartoon giant panda with a red face and a tuft of hair on her head. Xiao Chuan is modeled after the adored giant panda Qi Yi, who went viral around the world for her "leg-hugging" video. Qi Yi was born on 1st July, 2016, and hence got the name (Qiyi sounds like 1st July in Chinese). With a tuft of hair like a Wi-Fi antenna on top of her head, the energetic little girl loves to chase after keepers and hug their legs, which is its claim to fame.

What have giant pandas been through over

the past 8 million years? Did they meet the dinosaurs? How can such cute and cuddly giant pandas chew and swallow tough, hard bamboo? How can giant pandas that have survived the end of the glacial era still be vulnerable to extinction? How much have generations of panda conservationists sacrificed to protect them, and what have they achieved? What can you do to support ecological conservation?

Xiao Chuan will lead visitors step by step through all the exhibitions of the Giant Panda Museum to introduce them to the world of the giant pandas and learn about their evolutionary history, physiological characteristics, anecdotes, achievements in scientific research and the influence of giant pandas in today's world.

As you gain knowledge about giant pandas, your love for them will grow; then you need to know them better. So, let's open this book, enter this unique museum on paper, and follow Xiao Chuan to enjoy a fun tour around the world of giant pandas!

的"奇一"生性活泼，最爱追着饲养员奶爸抱大腿，这一抱就抱成了网络明星。

800万年的漫长岁月里，大熊猫们经历了什么？它们和恐龙见过面吗？又韧又硬的竹子，外表软萌的大熊猫怎么嚼得烂、咽得下？扛过了冰川纪大灭绝的彪悍大熊猫怎么还会面临生存危机？为了保护大熊猫，一代代"熊猫人"付出了多大的努力，取得了怎样的成果？普通人能力所能及地为生态保护做点儿什么……

大熊猫博物馆的每个展区里都有小川的身影，它会带着我们一步步进入大熊猫的世界，了解大熊猫的演化历史、生理特点，关于它们的奇闻轶事和最新科研成果，以及大熊猫在当今的影响力。

了解大熊猫，你更容易爱上它；爱上大熊猫，你更需要了解它。那就翻开这本书，走进这座独特的纸上博物馆，让我们跟着小川去畅游大熊猫的世界吧！

这本书里蕴藏着一个浩瀚的熊猫世界。

我们将一起穿越时空隧道,抵达 800 万年前。

从先祖始熊猫开始,大熊猫一族不断演化,成为如今的现生大熊猫。

历史不断前行,人类文明逐渐出现。族群交集的历史被记录在人类的文明中,大熊猫和人类一起面对了沧桑巨变。但是,随着人类的发展,大熊猫的生存空间越来越受到挤压。

大熊猫面临着生存危机!

幸运的是,人类很快意识到保护自然和环境的重要性。从建立自然保护区开始,经过几代人的努力,大熊猫终于渡过了生存危机。

保护大熊猫,也就保护了我们自己。

接下来,
我们将进入大熊猫的
专属地盘。
跟紧了哦,
一路上会有好多惊喜和意外。
嘿嘿!

出发
Go!

A vast world of pandas will unfold in this book.

Let's travel through time to 8 million years ago.

The giant panda family evolved from the very beginning as *Ailurarctos lufengensis*, to the modern-day giant panda.

As time rolls on, human civilization emerged. The history of the intermingling of communities was recorded by human civilization in which giant pandas and humans have faced great changes together. However, human beings' burgeoning development has chipped away at the living space of giant pandas.

That has put their survival under tremendous threat!

Fortunately, their human counterparts soon realized the significance of protecting nature. After several generations of efforts starting from the establishment of nature reserves, giant pandas have finally survived the crisis.

Protecting giant pandas is equivalent to protecting ourselves.

Next,
We will enter
the "exclusive territory"
of the giant pandas.
Stay close.
There will be many surprises along the way.
Hey, hey!

大熊猫博物馆入口
时光隧道

The Giant Panda Museum entrance
The Time Tunnel

Chapter 1　第一章

>> Tracing Origins
熊猫前传

黑白映画，创生物圈之亘古传奇；
桑海同源，承大自然的旷世遗珍。

In black and white, you are an ancient legend in ecosystems; born millions of years ago, you are an epic treasure hidden in nature.

Evolution History of the Giant Panda
大熊猫的演化史

8 Million Years Ago

每一种生物的诞生、演化，都有其自身的规律。

为什么大熊猫被称为"活化石"？它已经在地球上生存了至少 800 万年，在漫长而严酷的生存竞争和自然选择中，与大熊猫同期生活的动物们大多已灭绝，比如剑齿象、巨猿等，大熊猫却神奇地生存了下来，并保持了原有的古老特征。

大熊猫的祖先长什么样子？它们的体形、食性和生活区域发生过什么变化……古生物学家们对多年来在各地发现的化石进行研究，逐渐勾勒出大熊猫的演化轨迹。

The birth and evolution of every living thing has its own rules.

Why is the giant panda called a "living fossil"? The reason is as follows: It has survived on earth for at least 8 million years, over the long course of which giant pandas not only miraculously survived the harsh competition and natural selection but also retained their original ancient features while most of other animals such as *Stegodon* and *Gigantopithecus* that lived at the same time with the species ended up extinct.

What did the ancestors of the giant panda look like? What changes have taken place in their body shape, dietary preference and living areas... Paleontologists have studied fossils found in various places over the years and gradually outlined the evolutionary trajectory of the species.

>> 距今 800 万年前

食肉为主的始熊猫
>> Carnivorous *Ailurarctos*

始熊猫化石首次发现地

The places where fossils of *Ailurarctos* were first found

始熊猫化石最先发现于我国云南省禄丰和元谋两地。

20 世纪 70 年代，在离云南省禄丰县县城 9 千米的石灰坝村，采煤工人在一个褐煤矿井作业时发现了大批脊椎动物化石，这引起了中国科学院古脊椎动物与古人类研究所的注意。研究所在 1978 年派工作人员前往调查。此后，研究人员在煤矿工人的协助下，在这个地区进行了多年的发掘，获得了大批古猿化石和一些稀有的哺乳动物化石，其中就包括禄丰始熊猫化石。研究人员根据化石的生物地层时序判断，禄丰始熊猫生活在距今 800 万年前。

就在禄丰始熊猫化石被发现后不久，研究人员又在云南省元谋县发现了元谋始熊猫化石。元谋始熊猫化石标本的基本形态与禄丰始熊猫接近，但个体比禄丰始熊猫小，进化水平稍高，它的生活年代比禄丰始熊猫稍晚，大约在距今 700 万年前。

禄丰始熊猫和元谋始熊猫的发现证明了大熊猫的祖先生活在中国。从始熊猫化石的颌骨及牙齿结构看，始熊猫个子较矮小，只有现生大熊猫的约三分之一大小。

Ailurarctos fossils were first found in Lufeng and Yuanmou in Yunnan Province, China.

In the 1970s in Shihuiba Village, 9 kilometers from the county seat of Lufeng County, Yunnan Province, coal miners discovered a large quantity of vertebrate fossils in a lignite mine, which grabbed the attention of the Institute of Vertebrate Paleontology and Paleoanthropology of the Chinese Academy of Sciences (IVCAP). The Institute sent staff to investigate in 1978. Since then, the researchers, with the assistance of the coal miners, have conducted excavations in the area for many years, and obtained many fossils of ancient apes and other rare mammals, including those from the *Ailurarctos lufengensis*. According to the biostratigraphic chronology of the fossils, the researchers concluded that the *A. lufengensis* lived about 8 million years ago.

始熊猫（复原图）
Ailurarctos (restored image)

古植物学家从始熊猫的化石层位提取的大量孢粉显示，这个植物组合中并没有竹子。这可能是因为当时竹子的生长范围还很小，也与始熊猫的牙齿仅仅具备了食竹的雏形构造相吻合，同时说明了始熊猫并不以竹为生，还是以食肉为主。

800万年前，我国的西部地区到处都是低山河谷，还分布着大片的原始森林。进入新生代以来，随着印度洋板块的挤压及青藏高原的抬升，原来的地形、地貌发生了沧海桑田的变化。部分较低的区域变成了沼泽和丘陵相间的湿地，原来的森林逐渐演变成为灌丛和竹林。与此同时，生活在这一带的古食肉类动物逐渐适应了新的环境，大熊猫的直系祖先——始熊猫便生活在这样的环境下。

Shortly after this discovery, researchers discovered the *Ailurarctos yuanmouensis* fossils in Yuanmou County, Yunnan Province. The basic morphology of the *A. yuanmouensis* fossil is similar to that of *A. lufengensis*. Although smaller, the newly discovered fossils showed a slightly higher level of evolution. The species lived in a later age than the *A. lufengensis*, about 7 million years ago.

The discovery of these two fossils proves that the ancestors of the giant panda lived in China. In terms of the jawbone and tooth structure of the *Ailurarctos* fossils, the original species was short, about one-third the size of modern-day giant pandas.

A large amount of sporopollen extracted by paleobotanists from the fossilized soil layers of the *Ailurarctos* meant that there was no bamboo in this plant assemblage. This may possibly be due to the fact that bamboo was dispersed over a small area, which also coincided with the fact that the *Ailurarctos* implies the beginning of a bamboo-eating diet. The results also revealed that the *Ailurarctos* did not live on bamboo, but was still mainly carnivorous.

8 million years ago, the western part of China featured low mountains, valleys, as well as a large swath of primitive forests. Since the Cenozoic Era, with the extrusion of the Indian Ocean plate and the uplift of the Qinghai-Tibetan Plateau, considerable changes took place to the original topography and geomorphology. Some of the lower areas turned into wetlands between swamps and hills, and the original forests gradually evolved into thickets and bamboo forests. Meanwhile, the ancient carnivores that lived in this area gradually adapted to the new environment, and the *Ailurarctos*, the ancestor of the giant panda, lived in such an environment.

始熊猫骨架复原模型
Restored Skeleton Model of *Ailurarctos*

容我骄傲一下：我们成为"国宝"可不仅仅是因为萌，大熊猫一族历史悠久，在地球上已经生存了至少800万年！

我们的祖先始熊猫是一种以食肉为主的动物，臼齿小，体形也比较小。到了早更新世早期，始熊猫演化成了小种大熊猫，体形也变大了。又过了几十万年，大熊猫武陵山亚种出现了，它的个头比现在的大熊猫小一些。到了中更新世，巴氏大熊猫出现了，它比现在的大熊猫体形还要大一些。那时候，我们生活的范围包括了黄河、长江和珠江流域，在越南和缅甸地区也能看到我们的身影。到了全新世，地球的寒冷季逐渐过去了，我们才变成了现在的模样。

>>

Please allow us to take pride in one thing: we are regarded as the Chinese national treasure not only for our adorable appearance but also for our long history surviving for at least 8 million years.

Our ancestor, known as *Ailurarctos*, was mainly carnivorous, with small molars and a relatively small body. In the early stage of Early Pleistocene, the original species evolved into *Ailuropoda microta*, which was also larger. After a few hundred thousand years, *Ailuropoda melanoleuca wulingshanensis* appeared, which was a little smaller than its current counterparts. By the Middle Pleistocene, *Ailuropoda melanoleuca baconi* came into being, which was a bit larger than us today. Back then, we inhabited basins of the Yellow River, the Changjiang River, and the Pearl River, and left our footprints in Vietnam and Myanmar as well. It was in the Holocene, when the earth's vast cooling gradually died down, that we became what we are now.

Hi, I'm Xiao Chuan.

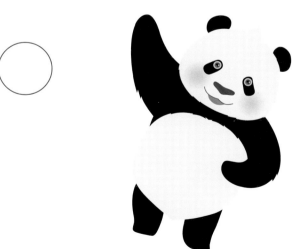

>>

Chengdu Giant Panda Museum Exhibition Hall 1 Tracing Origins

成都大熊猫博物馆 第一展厅 熊猫前传

科学家是如何判定化石具体年龄的?
How do scientists determine the specific age of fossils?

有用的知识增加了

在考古工作中,科学家常用碳-14年代测定法来确定古生物化石的年代。碳-14是一种放射性同位素,会随时间发生衰变。生物活着的时候,由于呼吸作用,一直与外界发生着碳循环,从而持续地从二氧化碳中吸收碳-14,体内的碳-14含量始终保持着一个恒定值。生物死亡后,因呼吸停止不再参与碳循环,体内碳-14的含量就会因衰变而减少。碳-14的半衰期为5730年,也就是说,过了5730年,碳-14就会有一半发生衰变。

因此,我们可以通过测量样品中碳-14的含量推算出古生物的生存年代。1947年,芝加哥大学的威拉德·利比因发明了碳-14年代测定法而获得诺贝尔化学奖。从此,考古学、地质学和水文学中令人头疼的年代确定问题得到了革命性的突破。

采用碳-14年代测定法能测得的相对准确的年代一般在6万年以内。用半衰期更长的放射性元素可以测定更久远的年代。比如,铀的三种天然同位素的半衰期:铀-238为45.1亿年,铀-235为7亿年,铀-234为24.7万年。

In archaeological work, scientists often use carbon-14 dating to determine the age of ancient fossils. Carbon-14 is a radioactive isotope that decays over time. When organisms are alive, they are in a carbon cycle with the outside world due to respiration, thus continuously absorbing carbon-14 from carbon dioxide, which means the amount of carbon-14 in the body always maintains a constant value. Once the organism dies, however, it ceases to absorb carbon-14, so that the amount of the radiocarbon in its tissues steadily decreases due to decay. Carbon-14 has a half-life of 5,730 years - i.e., half the amount of the radioisotope present at any given time will undergo spontaneous disintegration during the succeeding 5,730 years.

Therefore, an estimate of the date at which an organism died can be made by measuring the amount of its residual radiocarbon. In 1947, Willard Libby of the University of Chicago was awarded the Nobel Prize in Chemistry for his invention of carbon-14 dating. Since then, revolutionary breakthroughs have been made in dating technologies in archaeology, geology, and hydrology, by which scientists have long been plagued.

The relatively accurate dates that can be obtained by the measure are generally within 60,000 years. For even earlier periods, radioactive elements with longer half-lives can be applied. For example, the half-lives of the 3 natural isotopes of uranium are 4.51 billion years for U-238, 700 million years for U-235, and 247,000 years for U-234.

巴氏大熊猫牙齿化石
Tooth fossils of *Ailuropoda melanoleuca baconi*

开始吃竹子的小种大熊猫
Ailuropoda microta Adapts to a Bamboo Diet

>> 距今 250 万年前

小种大熊猫虽然是在始熊猫之后出现的,但是研究人员发掘出小种大熊猫化石并为之命名却是在发现始熊猫之前。

广西柳州有一个叫覃秀怀的农民,他常听父辈讲家乡岩洞里的泥土营养丰富,可以拿来当肥料,还听说能在岩泥里捡到值钱的"龙骨"。一天,他去岩洞里挖岩泥,果真找到了很多"龙骨"。1956年12月的一天,他挑着一担"龙骨"去市场上售卖。洛满人民银行营业所的主任韦跃社从担子里拿起一件像人的下颌骨的"龙骨"掂了掂分量,告诉覃秀怀这些"龙骨"可能有科学研究价值,建议他捐赠给政府。

这块"龙骨"被交到了正在广西南宁华南洞穴考察的古生物学家裴文中教授手中。经过研究,它正是考察队在寻找的巨猿下颌骨化石!

从1957年~1964年,考察队在覃秀怀家乡的岩洞里共发掘出脊椎动物化石数千件,其中就包括了小种大熊猫化石。至此,小种大熊猫开始被人们所认识和了解。因其个子矮小等相关特征,裴教授给这种大熊猫起了个学名:小种大熊猫。

小种大熊猫生活在更新世早期,距今约250万年前。它比始熊猫块头大多了,约有现生大熊猫的一半大小,有比较粗糙宽大的臼齿。它们已经开始吃竹子,但是对竹子的需求量还不高,仍然处于杂食阶段。

小种大熊猫时期,大熊猫家族已从云贵高原辐射分布至我国的华东、华南乃至西北内陆等地区。

During the era of the *A. microta*, the species mainly inhabited Yunnan-Guizhou Plateau before spreading to eastern and southern China and even the northwestern interior of China.

2.5 Million Years Ago

Although *Ailuropoda microta* appeared after the *Ailurarctos*, in the fossil record researchers unearthed the *A. microta* fossils and named them before the discovery of *Ailurarctos*.

A farmer named Qin Xiuhuai from Liuzhou, Guangxi Zhuang Autonomous Region, often heard his parents saying that the soil in the caverns of his hometown was rich in nutrients and could be used as fertilizer, and that he could pick up valuable "loong bones" in the rocky mud. One day, he went to the cavern for some of that mud and happened across piles of "loong bones". In December 1956, he carried a load of "loong bones" to sell at the market. The director of the sales office of the People's Bank of Luoman, Wei Yueshe, picked up a piece that resembled a human jawbone from the load, weighed it in his hands, and told Qin Xiuhuai that these bones might shed light on scientific research, and suggested that he should donate them to the government.

This piece was handed over to Prof. Pei Wenzhong, a paleontologist who was on an expedition to the South China Cave in Nanning, Guangxi Zhuang Autonomous Region. After investigating, it was determined that this was the mandible of *Giantopithecus* that the team was searching for!

From 1957 to 1964, the expedition team in the caverns of Qin Xiuhuai's hometown unearthed thousands of vertebrate fossils, including those of the *Ailuropoda microta*. From then on, the species began to be recognized and understood. Because of its small size and other related characteristics, Prof. Pei gave this kind of giant panda a scientific name: *Ailuropoda microta*.

The *A. microta* lived in the early Pleistocene, about 2.5 million years ago. They were bigger than *Ailurarctos*, about half the size of modern-day giant panda, and had rough and wide molars. At that time, they had begun to eat bamboo, though in small amounts, suggesting they were still omnivorous.

小种大熊猫化石发现地

Discovery site of the *Ailuropoda microta* fossils

小种大熊猫（复原图）
Ailuropoda microta (restored image)

大熊猫的化石是这样形成的
>> How Giant Panda Fossils Were Formed

没有一块化石是简简单单就形成的。化石是研究地球历史、生物发展史的重要见证。远古生物能成为化石并被人们发现，那机会真算是万中无一。来看看一块大熊猫化石要经历怎样的机缘巧合才可能出现在我们面前吧。

No fossil forms overnight. Fossils are key witnesses to the history of the Earth and the development of living creatures. It is a one in a million chance that an ancient creature will become a fossil and be discovered. Let's take a look at what stages a giant panda fossil has to go through before it can appear in front of us.

悠闲的生活

大熊猫悠闲地生活在竹林里，竹笋、竹叶和竹竿都是它们喜欢的食物。

A Leisurely Life

Giant pandas lead a care-free life in the bamboo forest, where bamboo shoots, leaves, and stems are their favorite food.

01

死亡来临

在自然规律的作用下，某一天，因衰老或其他原因，它们走向了死亡。

Looming Death

Based on the law of nature, they will meet their death one day due to aging or other reasons.

02

尸骨仅存

多数情况下，大熊猫的尸体暴露于地表，会完全腐败，不会形成化石；但有时它们的尸体也会被流水带到低洼处，软组织腐败，只剩下骨骼。

Remaining Carcasses

In most cases, the bodies of giant pandas are exposed on the surface of the ground and decay completely without fossilizing. However, sometimes their corpses are carried by running water to low-lying areas, where the soft tissues decay and only the bones remain.

03

深埋地下

泥沙将尸骨掩埋，上面的沉积物越来越厚，地下水中的矿物质进入骨骼空隙沉淀下来，甚至形成重结晶，日久天长就成了化石。

Buried Deep Underground

Silt buries the bones, on top of which the sediment becomes increasingly thicker. Then, minerals in the groundwater, which permeate bond voids, precipitate, and even recrystallize, becoming fossilized over time.

04

野外发掘

又过了很长很长时间，地质变化、自然风化作用或人类活动使大熊猫的化石暴露于地表，与人们相遇，成为古生物学家发掘、采集和科学研究的对象。

Wild Excavation

After a long period of time, geological changes, natural weathering or human activities expose the fossils of giant pandas on the surface of the ground, which will be discovered and become the object of paleontologists' excavation, collection, and scientific research.

05

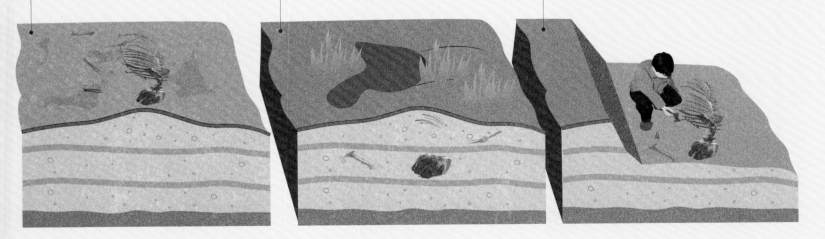

以竹为主食的大熊猫武陵山亚种　　>> 距今 180 万年前
>> *Ailuropoda melanoleuca wulingshanensis*
Eating Bamboo Exclusively

　　1978 年，中国科学院古脊椎动物与古人类研究所和湖南省博物馆组成的联合考察队在湖南省保靖县洞泡山的一个石灰岩洞穴内发现了大熊猫化石。研究人员将这种大熊猫化石与在广西柳州笔架山和湖北建始龙骨洞发掘的大熊猫化石对比，发现它们非常相似。因为都被发现于武陵山地区，人们将它们称为武陵山大熊猫。

　　武陵山大熊猫的体形比小种大熊猫大，比现生大熊猫稍小。它们生活在距今 180 万年前，已经以竹子为主食。

　　根据化石特征，多数研究者认为武陵山大熊猫是小种大熊猫与巴氏大熊猫之间的一个过渡成员，属于一个新的亚种——大熊猫武陵山亚种；但也有研究者认为，武陵山大熊猫应该被视为大熊猫演化过程中的一个独立的种。

湖南省
Hunan Province
● 保靖 Baojing

大熊猫武陵山亚种化石发现地

Discovery Site of the *Ailuropoda melanoleuca wulingshanensis* fossils

大熊猫武陵山亚种右上颌骨化石模型
Right maxilla fossil replica of *Ailuropoda melanoleuca wulingshanensis*

1.8 Million Years Ago

In 1978, an expedition team jointly organized by the Institute of Vertebrate Paleontology and Paleoanthropology of the Chinese Academy of Sciences and the Hunan Museum found the giant panda fossils in a limestone cave in Dongpao Mountain, Baojing County, Hunan Province. Researchers compared the giant panda fossils with those excavated at Bijia Mountain in Liuzhou, Guangxi Province and Longgu Cave in Jianshi, Hubei Province, and found that they belonged to the same group. As the area where those fossils were found is part of the Wuling Mountain, this kind of giant panda was named the Wuling Mountain giant panda.

The Wuling Mountain giant panda was larger than the *Ailuropoda microta*, although slightly smaller than the living ones. They lived 1.8 million years ago and mainly fed on bamboo.

Based on fossil characteristics, most researchers believe that the Wuling Mountain giant panda is a transitional member between the *Ailuropoda microta* and *Ailuropoda melanoleuca baconi*, belonging to a new subspecies — *Ailuropoda melanoleuca wulingshanensis*. However, there are also researchers who believe that the Wuling Mountain giant panda should be regarded as an independent species in the evolution of the giant panda.

> 在中国南方的更新世化石层里，常常见到长臂猿、大熊猫和巨猿的化石埋藏在一起。

> Gibbon fossils are often seen together with those of giant pandas and the *Giantopithecus* in the Pleistocene fossil layer in southern China.

大熊猫武陵山亚种（复原图）
Ailuropoda melanoleuca wulingshanensis
(restored image)

大块头的巴氏大熊猫
>> Large *Ailuropoda melanoleuca baconi*

>> 距今 75 万年前

巴氏大熊猫化石最早是在缅甸被发现的。1915 年，在缅甸摩谷城鲁比矿区的一个洞穴里，采矿工人发现了一些哺乳动物化石。其中有一件完整的上颌骨，上面还保存有多枚牙齿。研究者认为，这件标本的形态特征很像生活在中国西部山区的现生大熊猫，但又有区别。考虑到化石埋藏于洞穴堆积物中，时代较早，就给它单独建立了一个新种，叫巴氏大熊猫。

后来，中国研究人员在国内的四川、广西、广东、湖南、湖北、云南、海南、福建、江苏、浙江、河北、山西、河南、陕西、甘肃等省（自治区）的上千个洞穴中相继发掘出了巴氏大熊猫化石。巴氏大熊猫化石在与中国相邻的泰国、老挝和越南等国也有发现。

研究推断，巴氏大熊猫生活在距今 75 万年前。因为经历了多次气候冷暖交替，它们的体形进一步增大，比现生大熊猫还要大九分之一到八分之一。体形和食量的增大使巴氏大熊猫进一步向珠江、长江和黄河流域温暖湿润的季风区扩散。在更新世中晚期，它们的分布范围达到历史上的最大，这也是大熊猫家族史上最为繁盛的时期。

巴氏大熊猫（复原图）
Ailuropoda melanoleuca baconi (restored image)

750,000 Years Ago

The *Ailuropoda melanoleuca baconi* fossils were first discovered in Burma (present Myanmar). In 1915, miners found a number of fossilized mammals in a cave in the Ruby mining area of Mogok, Burma. Among them was a complete maxilla with several teeth preserved. The specimen, researchers theorized, resembled that of a living giant panda that inhabited the mountains of western China, but there were differences. Considering that the fossil was buried in cave deposits of an earlier age, scientists defined the new species as the *A.m.baconi*.

In the following years, Chinese researchers successively unearthed *A.m.baconi* fossils in thousands of caves in Sichuan, Guangxi, Guangdong, Hunan, Hubei, Yunnan, Hainan, Fujian, Jiangsu, Zhejiang, Hebei, Shanxi, Henan, Shaanxi, Gansu, and other provinces (or autonomous regions) in China. Fossils of the species have also been found in China's neighboring countries such as Thailand, Laos, and Vietnam.

According to relevant studies, the species lived 750,000 years ago. Because of the alternating warm and cold climates, they were one-ninth to one-eighth larger than modern-day giant pandas. The increase in size and food intake enabled the pandas to travel further into the warmer and more humid monsoon areas of the Pearl River, Changjiang River, and Yellow River Basins. The middle and late Pleistocene saw their widest presence in history, which also marked the most thriving period in the history of the giant panda family.

摩谷城 Mogok

缅甸
Burma (present Myanmar)

巴氏大熊猫
化石首次发现地

Where *Ailuropoda melanoleuca baconi* fossils were first found

巴氏大熊猫的化石在中国的很多地区,以及缅甸、泰国、老挝和越南等国都有发现。

Fossils of the *Ailuropoda melanoleuca baconi* were found in many parts of China, as well as in Burma, Thailand, Laos, and Vietnam.

远古大熊猫的"朋友圈"
>> "The Friend Circle" of Ancient Giant Pandas

在距今258万年前至1.2万年前的更新世，秦岭—淮河以南气候适宜、森林广布、食物丰富，哺乳动物繁盛，动物数量非常多。这里繁衍着一个动物群，生活着包括大熊猫、东方剑齿象、剑齿虎、巨猿、三趾马、巨貘、猩猩、纳玛象、中国犀牛、水牛、猕猴、金丝猴、长臂猿、野猪、豪猪、亚洲黑熊等在内的多种动物，地质学家称它们为大熊猫—剑齿象动物群。

In the Pleistocene period from 2.58 million to 12,000 years ago, south of the Qinling-Huaihe region was endowed with a favorable climate, vast forests, and abundant food, which was conducive to mammalian reproduction, and thus an enormous variety of animals inhabited the area, including giant pandas, *Stegodon orientalis*, *Smilodon*, *Giantopithecus*, *Hipparion*, *Megatapirus augustus*, gorillas, *Palaeoloxodon namadicus*, Chinese rhinos, buffalos, macaques, snub-nosed monkeys, gibbons, wild boars, porcupines, *Ursus thibetanus* and more. They were referred to by geologists as the *Ailuropoda-Stegodon* fauna.

剑齿象（已灭绝）

剑齿象长着长长的象牙，体重约12吨，植食为主，生活在热带及亚热带沼泽和河边的温暖地带，分布在亚洲和非洲。东方剑齿象是大熊猫——剑齿象动物群的重要成员。

Stegodon (Extinct)

The *Stegodon* had long tusks and weighed about 12 tons. The herbivore lived in tropical and subtropical swamps and warm riverside areas in Asia and Africa. The *Stegodon orientalis* is an essential member of the *Ailuropoda-Stegodon* fauna.

剑齿虎（已灭绝）

剑齿虎长着标志性的锋利牙齿，但它并不是现代老虎、狮子或家猫的祖先，而是猫科动物的另外一支。它重达200千克，猎捕对象是大型植食动物，曾广泛分布在亚洲、欧洲和美洲。

Smilodon (Extinct)

The *Smilodon*, with its iconic razor-sharp teeth, is not the ancestor of the modern tigers, lions or domestic cats, but belongs to a different branch of the feline family. The species weighed up to 200 kilograms and preyed on large herbivores. It once roamed across Asia, Europe, and the Americas.

三趾马（已灭绝）

　　三趾马是在草原古马的基础上进化而来的。它比现代马小，前后肢均有三趾。它体长约1.2米、肩高约1米，以植物为食。三趾马最初生活在森林中，后来来到草原上繁衍生息。它曾分布在欧洲、亚洲、非洲和美洲。

Hipparion (Extinct)

　　Hipparion evolved from the ancient steppe horse. The species had body that measured at about 1.2 meters and a shoulder height of about 1 meter. It was smaller than the modern horse and had 3 toes on both its front and hind limbs. The herbivore initially lived in the forest, and later traveled to the grasslands to reproduce. It once lived in Europe, Asia, Africa, and the Americas.

巨猿（已灭绝）

　　巨猿是世界上存在过的最大的猿，站立身高近3米，重达500千克。它是素食者，最喜欢的食物是竹子，偶尔吃树叶和果实。它生活在森林里，曾分布在中国、印度及越南。

Giantopithecus (Extinct)

　　Giantopithecus was the largest ape that ever existed, standing nearly 3 meters tall and weighing 500 kilograms. It was an herbivore, whose favorite food was bamboo, and occasionally leaves and fruits. It lived in forests and was once found in China, India, and Vietnam.

栖息地退缩的现生大熊猫
Ailuropoda melanoleuca Faced with a Shrinking Habitat

>> 距今 1.2 万年前

第四次冰期的高峰到来以后，气候变得更加寒冷，加上喜马拉雅造山运动，地球的自然环境发生了剧烈变化。进入全新世后，很多动物都销声匿迹了，就连剑齿虎、剑齿象、中国犀牛、巨猿等威武雄壮的猛兽都没能逃过灭绝的厄运。虽然巴氏大熊猫种群也逐渐衰退，数量骤减，但大熊猫这一物种却神奇地生存了下来，并延续至今。

我们通常把进入全新世后，距今 1.2 万年前以来的大熊猫称为现生大熊猫。随着最后一次冰期结束，气候回暖。在人类的史前时期，大熊猫在中国的山西、河南、湖北、湖南、贵州和云南等地均有分布，但自从进入铁器时代以来，随着人口增加和人类生活区域的扩大，大熊猫进一步向西部偏远山区退缩。目前，野生大熊猫仅生活在四川、陕西、甘肃三省的秦岭、岷山、邛崃山、大相岭、小相岭及凉山这六大山系的森林中。

目前野生大熊猫的分布范围
Current Distribution Scope of Wild Giant Pandas

12,000 Years Ago

After the peak of the Fourth Ice Age, it was getting colder, and coupled with the Himalayan orogeny, the Earth's natural environment underwent drastic changes. Since the Holocene, many animals have gone extinct, and even giant ones such as the *Smilodon*, *Stegodon*, Chinese rhinos, and *Giantopithecus* did not escape their fate. Although the population of the *Ailuropoda melanoleuca baconi* plummeted, the species miraculously survived, and their descendants still continue to this day.

Ailuropoda melanoleuca usually refer to those born since the Holocene, 12,000 years ago. As the last ice age came to an end, the climate warmed up. During the prehistoric period of mankind, giant pandas were found in Shanxi, Henan, Hubei, Hunan, Guizhou, and Yunnan provinces in China, but ever since the Iron Age, the increase in population and the expansion of human living space have squeezed giant pandas out of their habitat further to the remote mountainous areas in the west. At present, wild giant pandas only live in the forests of the 6 mountain systems of Qinling, Min, Qionglai, Daxiangling, Xiaoxiangling, and Liang in Sichuan, Shaanxi, and Gansu provinces.

在人类史前时期，大熊猫在中国的山西、河南、湖北、湖南、贵州和云南等地均有分布。

During human prehistory, giant pandas were found in Shanxi, Henan, Hubei, Hunan, Guizhou, and Yunnan provinces in China.

现生大熊猫
Ailuropoda melanoleuca

大熊猫的体形变了又变
>> Changing Physiology of Giant Pandas

大熊猫的体形从古至今经历了从小变大再变小的过程。

Giant pandas have grown from small to large and then small again since ancient times.

现生大熊猫
Ailuropoda melanoleuca

1.2 万年前至今 12,000 years ago to present

巴氏大熊猫
Ailuropoda melanoleuca baconi

75 万年前 750,000 years ago

大熊猫武陵山亚种
Ailuropoda melanoleuca wulingshanensis

180 万年前 1.8 million years ago

小种大熊猫
Ailuropoda microta

250 万年前 2.5 million years ago

始熊猫
Ailurarctos

800 万年前 8 million years ago

Chapter 2 第二章
Among the Pandas
竹林隐士

物我相融，独爱天府丛林；
憨态可掬，固守神州乐土。

Integrating into nature, an unique evolution, enthusiast of bamboo in habitats of abundance; charming and naive, a custodian of happiness in China.

经过数百万年的演化，我们的栖息地改变了，退出了食肉者行列，成为吃竹子的素食者，从此深居简出，从上古神兽变成与世无争的"竹林隐士"。

作为演化过程中的佼佼者，我们有着对自己来说最优化的采食方式和独特的消化系统，有适应环境的黑白相间的厚实皮毛，还有在野外交流相处、繁育后代的独特法则，这些特点使得我们能够在茂竹与清泉之间悠然自得地繁衍生息，经历冬夏、感受四季、度过千万年的沧海桑田，直到在今天与你相遇。

来看看我们为了生存做出的努力和改变吧！

》》 After millions of years of evolution, we are no longer carnivorous but bamboo-eating vegetarians due to our changes of habitat. Since then we have been living in deep seclusion, evolving from ancient mammals to "bamboo hermits".

As the best in the process of evolution, we boast an optimal feeding method, unique digestive system, thick black and white fur perfectly adapted to the environment, and distinctive ways of communicating and getting along with each other in the wild and reproducing our offspring. Those characteristics have enabled us to live and multiply in lush bamboo forests with clear springs. After passing through millions of years of twists and turns, we have made it to meet with you today.

Let's take a look at what we have done to survive!

嗨，我是小川。
Hi, I'm Xiao Chuan.

\>\>

Chengdu
Giant Panda
Museum
Exhibition Hall 2
Among the Pandas

成都大熊猫博物馆 第二展厅 竹林隐士

The Giant Panda Body

大熊猫的身体

① **头部：** 头骨较大，颧骨弓发达，矢状脊突出，有利于强大的咬肌附着。
② **眼眶：** 黑眼圈有吸收紫外线防护眼睛、威慑天敌以及个体识别的作用。
③ **牙齿：** 成年个体一般有40~42颗牙。
④ **皮毛：** 皮毛厚而粗，有良好的防潮、御寒作用。
⑤ **四肢：** 四肢粗壮有力，足以支撑壮硕的身躯。
⑥ **尾巴：** 扁平的白色尾巴，相对于庞大的身躯显得非常短小。
⑦ **伪拇指：** 前掌由桡侧籽骨特化而成"第六指"，可与其他五趾对握，便于抓握食物。
⑧ **脚掌：** 生长着密密的丛毛，可防止在潮湿的坡面行走时滑倒。
⑨ **爪子：** 前后掌上长着锋利的趾爪，便于爬到树上休息、玩耍和躲避敌害。

① **Head:** The skull is large, with a well-developed zygomatic arch and prominent sagittal ridges, which is conducive to the attachment of powerful biting muscles.
② **Eye sockets:** The dark circles around eyes can absorb ultraviolet rays to protect the eyes, deter natural enemies, and be identified.
③ **Teeth:** Adult individuals generally have 40 to 42 teeth.
④ **Fur:** Their thick and coarse fur is moist-proof and keeps themselves warm.
⑤ **Limbs:** Their limbs are strong enough to support their huge body.
⑥ **Tail:** Their flat white tail appears very short when compared to their huge body.
⑦ **Pseudo-thumb:** The giant panda has a greatly enlarged wrist bone, the radial sesamoid, that acts as a sixth digit, an opposable "thumb" to easily manipulate bamboo.
⑧ **Paws:** Dense tufts of hairs prevent them from slipping when they are walking on wet slopes.
⑨ **Claws:** Front and rear palms with sharp toes and claws make it easy for them to climb up trees to rest, play, and hide from enemies.

毛色：黑白配色最安全
Fur Color: Safest Black and White

　　大熊猫的四肢、肩带、耳朵、眼眶、鼻子部位的毛发均为黑色，其余部位的毛发为白色。这样的配色不仅好看，更重要的是能发挥保护色的作用。

　　大熊猫常年生活在光影斑驳的山林中，黑白相间的体色与白天阳光照射下山林中明暗相间的影像相似，从而能很好地隐蔽自己。如果在积雪覆盖的栖息地，它们身上白色的部分刚好与环境融为一体，而黑色部分则能帮助它们将自己藏在阴影里。

> **大熊猫的黑白色皮毛在野外有很好的保护和隐蔽作用。**
> The black and white fur of the giant panda acts as excellent camouflage for protection in the wild.

Giant pandas have black fur on their limbs, shoulder straps, ears, eye sockets and nose, and white fur on the rest of their body. Their pattern is not only lovely, but more importantly, it can play a protective role.

Giant pandas live in mountains and forests with mottled tree shadows all year round. As their black and white fur is similar to these shadows during the daytime, they can camouflage themselves well. In a snow-covered habitat, the white fur blends in perfectly with the environment, while the black helps them hide themselves in the dark.

有其他毛色的大熊猫吗?
>> Are There Giant Pandas with Other Fur Colors?

大熊猫的皮毛颜色通常是黑白相间的,生活在不同山系的大熊猫的皮毛颜色没有什么区别。人们仅发现过个别皮毛颜色异常的大熊猫。

棕白色大熊猫

从1985年到现在,人们10余次在秦岭一带发现棕白相间的大熊猫,其中最著名的是大熊猫"七仔",人称"巧克力熊"。

白色大熊猫

2019年,人们首次在野外记录到全球唯一一只白色大熊猫。据分析,这是一只"白化"个体。"白化"现象通常是由于动物个体基因突变导致体内无法正常合成黑色素而形成的。

Their fur color is usually black and white, and thus there is no difference between the fur colors of giant pandas living in different mountain ranges. Only a handful of giant pandas with abnormal fur color have been found.

Brown and White Giant Panda

From 1985 to the present, brown and white giant pandas have been spotted more than 10 times along the Qinling Mountains. The most famous one is the giant panda Qi Zai, also known as the Chocolate Bear.

White Giant Panda

In 2019, the world's only white giant panda was recorded in the wild for the first time and was determined to be an albinistic individual. Albinism is usually caused by a genetic mutation in an individual animal that prevents the body from synthesizing melanin normally.

眼斑：黑眼圈作用大
>> Dark Circles: Crucial Function

大熊猫的眼睛周围有一块黑色的眼斑，看上去就像两个黑眼圈。这是大熊猫在几百万年的演化中为了更好地生存进化而来的。黑眼圈不仅是大熊猫外表上的典型特征，对大熊猫的生存也具有重要作用。

防护作用： 黑色的眼圈能增加大熊猫眼睛周围皮毛吸收紫外线的面积，从而减少紫外线对它们眼睛的刺激。所以，黑眼圈对于大熊猫的作用类似于墨镜对于人类的作用。

威慑作用： 有的科学家认为，大熊猫与猛兽狭路相逢时，能通过大大的黑眼圈，用气势来吓跑对方。大熊猫的眼部几乎是全黑的，其他动物可能会误以为自己碰到了一个眼睛巨大的怪物，从而仓皇逃跑。

识别作用： 每只大熊猫的黑眼圈都有着独一无二的轮廓，就像人类的指纹一样。科学家认为，大熊猫之间可以通过黑眼圈来相互识别。

Giant pandas have dark circles around their eyes that look like they have not slept in a while. The feature has evolved over millions of years for better survival. It is not only a typical feature of their appearance, but also plays an important role in their survival.

Protection: The dark circles can increase the area of fur around their eyes to absorb ultraviolet rays, thus reducing accompanying stimulation on their eyes. Therefore, the role is similar to that of sunglasses for humans.

Deterrence: Some scientists believe that when giant pandas encounter predators, they can scare away the threat thanks to their dark circles because predators may mistake those almost completely "black eyes" as those of a monster, thus running away in a hurry.

Recognition: Each giant panda's dark circle has a unique outline, like a human fingerprint. Scientists believe that giant pandas can recognize each other through these circles.

皮毛：这一身防寒又防潮
>> Fur: Moist-proof and Warm

大熊猫的皮毛看上去柔软而光滑，实际上却非常厚密、硬直。它们的皮脂腺很发达，分泌出的油脂使被毛光滑，既有利于保温、防水，也对它们在竹林中穿行起到了保护作用。它们的被毛厚而粗，由针毛和绒毛组成，从毛发的显微结构看，其外层髓质比较厚，有良好的防寒防潮作用。这样的皮毛把大熊猫严严实实地包裹并保护起来，让它们可以在潮湿、严寒的自然环境里随意行动，即便在积雪覆盖大地的季节也能照样外出觅食，所以大熊猫并不需要冬眠来度过漫长的冬天。

Their fur looks soft and smooth but is actually dense and stiff. Their sebaceous glands are well developed and secrete oil to smooth out the coat, which is conducive to heat preservation and dispelling water, plus it also protects them when they are walking through bamboo forests. Their coat is thick and coarse due to guard hair and fine hair. In terms of the microstructure, its outer layer of medulla is thicker, which enables them to move freely in the wet, freezing environment by keeping them comfortably warm, and even in the severe cold season, they can still forage for food. That is why pandas do not need to hibernate during the lengthy winter.

幼年和成年大熊猫都会换毛。根据对圈养大熊猫的观察，大熊猫换毛最明显的时期是1.5~2岁时。

Both juvenile and adult giant pandas molt. According to the observation of captive giant pandas, the most obvious molting period is when they are 1.5 to 2 years old.

骨骼：胖归胖，爬树攀岩轻松上
Bones: Chubby but Flexible

大熊猫的颈椎非常短，只有 7 节椎骨，这使得它们的头部转动灵活。这些颈椎骨也很粗壮，可以承受大熊猫的头部重量，从而让它们更容易觅食。

它们头骨的颧骨宽厚且向外扩展，矢状脊特别发达，可附着发达的咀嚼肌，这种生理结构是它们吃竹子的力学基础。

大熊猫四肢的小腿骨很粗壮，趾骨分明且前端锋利，这使得它们可以轻松地攀爬树木和岩石，也可以在不易攀爬的地形上行走。

Giant pandas have a very short cervical spine with only 7 vertebrae, which allows them to turn their heads quite far. These cervical vertebrae are also strong enough to support their heads, making it easier for them to forage for food.

The zygomatic arch of their skulls is wide, thick, and expanded, and the muscles attach to the strong sagittal crest toward the back of the skull, giving the jaw a great deal of leverage, which enables them to eat bamboo.

Their strong calf bones with well-defined, sharp toes allow them to climb trees and rocks with dexterity, and to walk on terrain that is not easily passable.

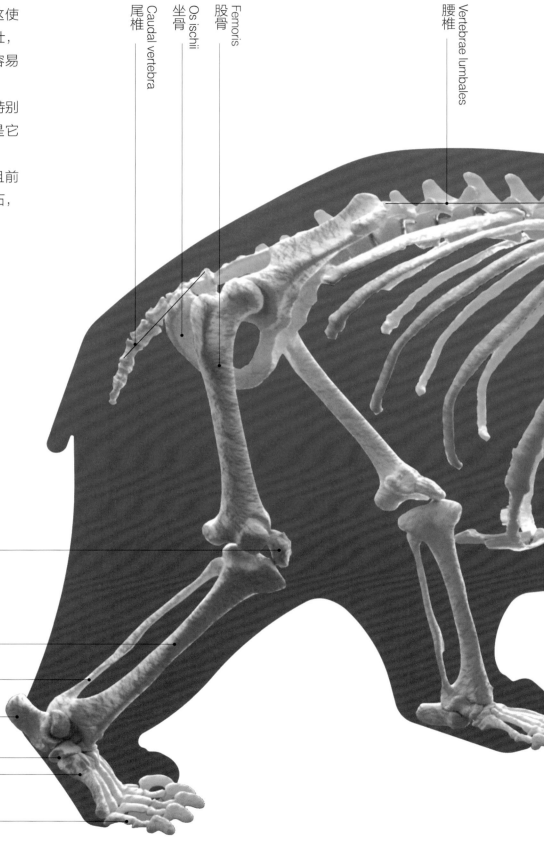

尾椎 Caudal vertebra
坐骨 Os ischii
股骨 Femoris
腰椎 Vertebrae lumbales
膝盖骨 Patella
胫骨 Tibia
腓骨 Fibula
跟骨 Calcaneus
跗骨 Ossa tarsi
跖骨 Ossa metatarsalia
趾骨 Ossa digitorum pedis

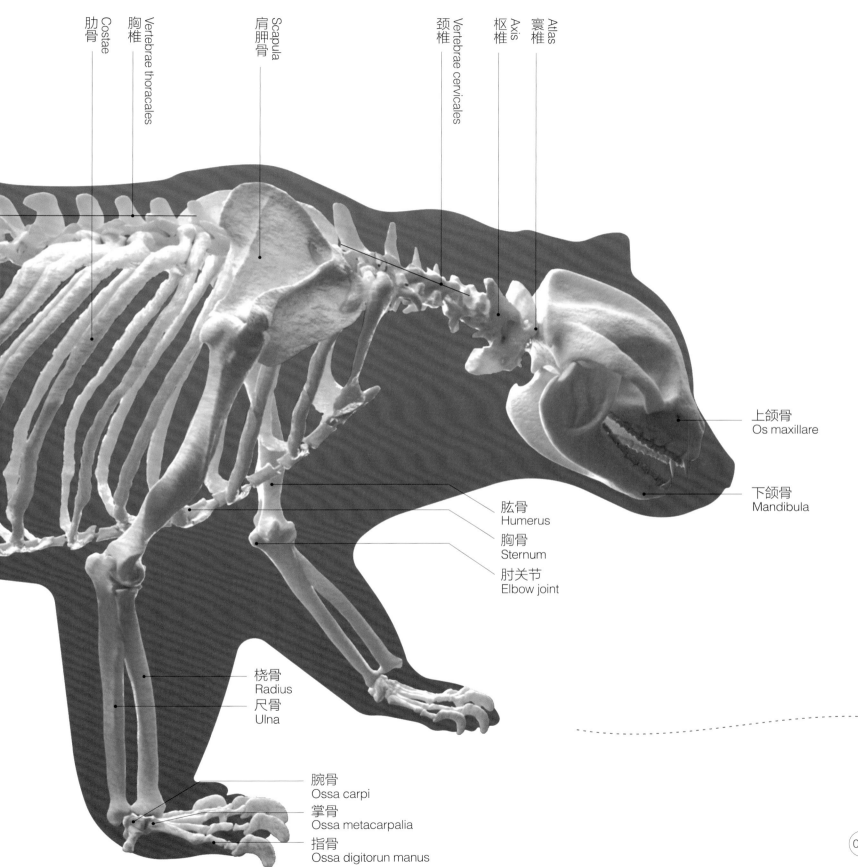

肌肉：外表小绅士，肉里肌肉男
Muscle: Gentlemanly Appearance but Strong

大熊猫圆乎乎、胖嘟嘟，看起来斯斯文文，甚至有些笨拙，但它们的肌肉其实非常发达。大熊猫的肌肉主要分布在腰、臀、肩胛和腿等几个部位，这让它们可以轻松地攀爬和行走。不过，跟大家认知里身形矫健的肌肉男不同，大熊猫虽然拥有壮硕的肌肉，但并不擅长奔跑，因为它们体重较大，肌肉形态并不适合长距离奔跑和快速移动。

The chubby giant panda looks gentle and even a bit clumsy, but they are actually very powerful. Their muscles are mainly distributed across the waist, buttocks, shoulder blades and legs, which allows them to climb and walk with ease. However, unlike athletic muscular men as assumed, giant pandas are not fast runners despite their strong muscles. Their heavy weight and musculature make them incapable of running long distance and fast movement.

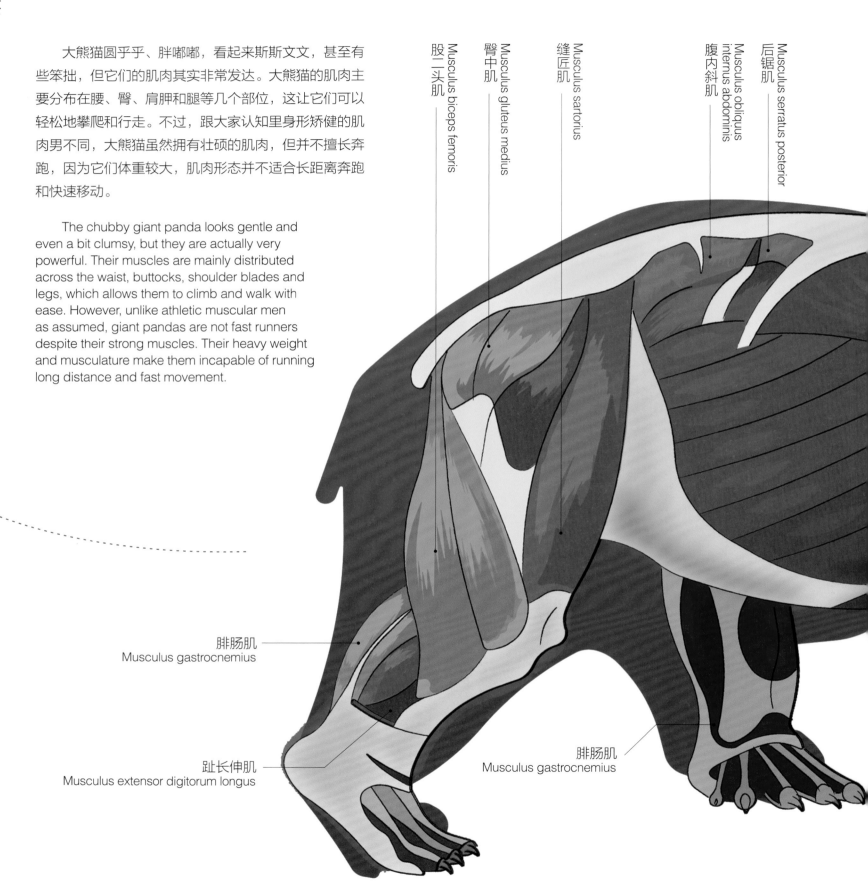

股二头肌 Musculus biceps femoris
臀中肌 Musculus gluteus medius
缝匠肌 Musculus sartorius
腹内斜肌 Musculus obliquus internus abdominis
后锯肌 Musculus serratus posterior
腓肠肌 Musculus gastrocnemius
趾长伸肌 Musculus extensor digitorum longus
腓肠肌 Musculus gastrocnemius

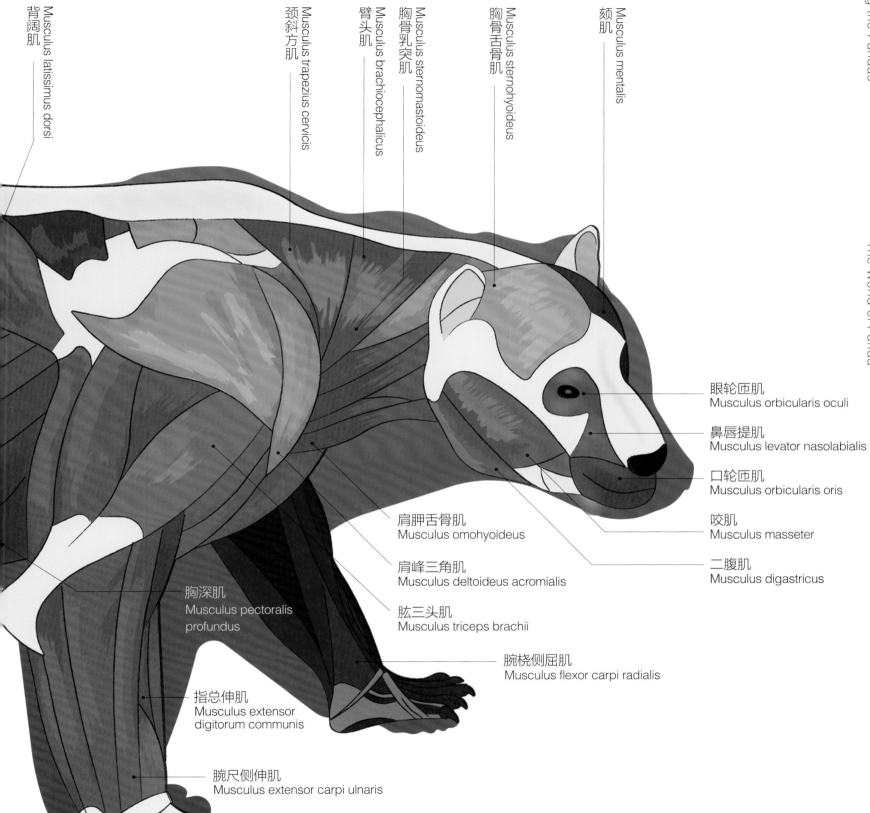

牙齿：最多能长出 42 颗
>> Teeth: 42 Maximum

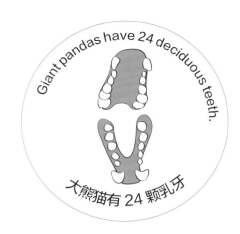

大熊猫出生后约 3 个月开始长牙；大约 5 个月时，乳牙长齐，有 24 颗；大约 8 个月时，开始进入换牙期；大约 1 岁半时，恒牙长齐。理论上大熊猫的牙齿为 42 颗，但有时个别牙齿不能萌出（就像人的智齿一样），所以实际上成年大熊猫的牙齿数量为 40~42 颗不等。

Giant pandas begin teething about 3 months after birth. At about 5 months, they have 24 deciduous teeth; At about 8 months, they begin to change their teeth, and when they are about 1.5 years old, their permanent teeth come in. Pandas usually have 42 teeth, but sometimes individual teeth do not erupt (like wisdom teeth in humans), so the actual number in adult pandas ranges from 40 to 42.

我吃素，但最好别惹我
>> Herbivores Not to Be Trifled With

别看大熊猫平时吃素，但它们发起怒来危险性极高，其咬合力在熊科动物中排第三。它们虽然个头比棕熊和北极熊小，但若按同等"重量级"来PK，它们的咬合力却远强于棕熊和北极熊。

Never underestimate giant pandas though they mainly feed on bamboo, they are extremely aggressive when threatened. Their bite force is the third strongest in Ursidae family. Giant pandas may be smaller than brown bears and polar bears, but their bite force is much stronger than these two in terms of the same heavy class.

熊科动物咬合力排名
Rank of Ursidae Bite Force

名称 Name	排名 Rank
北极熊 Polar bear	01
棕熊 Brown bear	02
大熊猫 Giant panda	03
马来熊 Sun bear	04
亚洲黑熊 Asiatic black bear	05
眼镜熊 Spectacled bear	06
美洲黑熊 American black bear	07

Senses and Language of the Giant Panda

大熊猫的感官和"语言"

嗅觉：谈恋爱，我们靠闻
>> Sense of Smell: Sniffing for Love

Giant pandas have lived in dense, dark bamboo forests for as long as they have existed, which has degraded their eyes, equivalent to 800 degrees of myopia in humans. Their sense of smell, however, is well developed. Scent marking is the most important way that they communicate between one another. They will apply urine, perianal gland secretions, or a mixture of the two to posts, stumps, walls, the ground, and places they often pass by to mark resources and territory. They will bob their heads with their mouths half-open as they make their marks. After that, they will peel off bark and leave scratches to attract other pandas' attention.

During the mating season, giant pandas also search for their mates mainly through their sense of smell. The scent markings of a female giant panda may indicate that she is ready for a relationship and allow male pandas to find her through these special signals. Interestingly, the reason why male pandas pee upside down is that they can urinate higher up, because the higher it is, the more mighty they seem, making female pandas to be much more interested. When love is not in the air, as soon as they get their first whiff of an unfamiliar giant panda, they will avoid each other because they all have their own territory.

大熊猫长期生活在茂密、阴暗的竹林中，视觉功能相对退化，相当于人的800度近视眼，但是嗅觉功能却非常发达。气味标记是大熊猫之间最主要的交流方式。它们会将尿液、肛周腺分泌物或二者的混合物涂在柱子、树桩、墙上、地上，以及它们经常经过的地方，用来标记资源和领地。它们做标记的时候，会晃动头部，嘴巴半张。做了标记以后，它们会用剥掉树皮、留下抓痕等方式来引起其他大熊猫的注意。

到了发情交配季节，大熊猫也主要通过嗅闻气味寻找配偶。雌性大熊猫的气味标记可能表示它已经做好谈恋爱的准备，雄性大熊猫可以通过这种特殊的"信号源"找到它。有趣的是，雄性大熊猫之所以倒立着尿尿，为的就是能够把尿液排到更高的地方。尿的位置越高，意味着雄性大熊猫的体形越大，就越能获得雌性大熊猫的青睐。在非恋爱季节，一闻到陌生大熊猫的气味，它们就会彼此避开，毕竟它们都有自己的领地。

成年大熊猫是一种独居动物，它们会通过做气味标记的方式来标记自己的领地，在发情期，也会通过气味标记来寻找配偶。

Adult pandas are solitary animals who mark their territory with scent markings, and during the rutting season, they will search for mates this way.

听觉和语言：吱吱叫咩咩叫，你听懂没？
Hearing and Communication: Can you Get the Squeaking and Baaing?

大熊猫也有自己的语言。它们的听觉比较灵敏，在不同时期、不同情况下它们都会通过叫声进行交流。这些叫声有的用于母子交流、有的用于找寻配偶、有的用于表达威慑或警告。

大熊猫宝宝才出生就能发出吱吱、哇哇、咕咕的叫声，而且声音比较尖锐。当它们饿了、身体不舒服或者受到了惊吓、想得到妈妈的安慰的时候，就会发出叫声，以吸引妈妈注意。

随着年龄增长，大熊猫能够发出的声音也越来越多。当它们受到惊吓的时候，会发出尖叫声。生气时，它们会发出怒吼声。而当它们感到舒服的时候，则会发出酥软的哼哼声。

到了谈恋爱的季节，雄性大熊猫会发出犬吠声。这是在向周围的雄性大熊猫发出警告："离我远点儿，别惹我！"也是在向雌性大熊猫传达爱意："看我有多强大！"

当一对大熊猫双双坠入爱河，雄性大熊猫会发出咩叫声，雌性大熊猫的叫声则会由咩叫声变为类似于鸟鸣的唧唧声。

Giant pandas have their own language. They have sensitive ears, and they communicate through their calls at different times and in different situations. Some of these calls are used exclusive between families, some are to find mates, and some are a warning and deterrent.

Panda cubs can squeak and make wah-wahs and coos with their high-pitched voices as soon as they are born. When they are hungry, unwell, or frightened and want their mothers to comfort them, they will cry to attract their mothers.

As they get older, they are able to make more sounds. When they are startled, they scream; when angry, they roar; and when they feel comfortable, they softly hum.

During the mating season, male pandas bark, a signal warning nearby male pandas, "Don't mess with me! Stay away!" and expressing desire for female pandas, "See how powerful I am!"

When a pair of pandas fall in love, the male panda will bleat, and the female panda's call will change from a bleat to a chirping sound kind of like a bird.

嗯叫：幼仔寻母
Hum: call for mother

歆叫：拒绝、厌烦
Snort: reject, annoy

哇哇叫：幼仔不适、饥饿
Bowwow: uncomfortable, hungry

尖叫：害怕、惊恐
Squeal: scared, frightened

犬叫，嗷叫：打斗、警告
Bark and Growl: fight, warn

咩叫：发情、求偶
Bleat: estrous, courtship

研究人员通过声谱特征分析，又参照它们的行为表现，已经解密了大熊猫的10余种叫声的含义。

Researchers have deciphered the meanings of more than 10 of the giant panda's calls by characterizing and associating their sound spectrum to their behavior.

吃竹子的食肉目动物
>> Carnivores Feeding on Bamboo

Giant pandas were originally carnivorous. Over the past few million years, with the changes in the environment and the advancement of human beings to the plains, gentle slopes and other zones suitable for reclamation and habitation, giant pandas have gradually retreated from their original habitat to the forests of the 6 mountain systems in western China, including the Qinling Mountains, Min Mountains and so on. These areas are rife with bamboo forests, and since bamboo undergoes asexual reproduction, its fast expansion can easily meet the feeding needs of pandas. Therefore, they made the best evolutionary choice by giving up meat and adapting to a bamboo diet, a drastic change.

Pandas gnawing on bamboo might be mistaken by many to be herbivores, or omnivores, while in fact, they are still categorized as carnivores. Diet was regarded as a major reference index of animal classification before, but that indicator alone cannot be accurate enough. Based on their digestive tract anatomy, physiological characteristics, and evolution, giant pandas are actually genuine carnivores. They evolved from carnivorous mammals with placentas over millions of years.

大熊猫原本是以吃肉为主的。几百万年来，随着大环境的变化和人类向平原、平缓的山坡等适宜开垦、适宜居住的地带的推进，大熊猫逐步向我国西部的秦岭、岷山等六大山系的森林中退缩。这些地带竹林众多，而且竹子属于无性繁殖，迭代速度快，能很好地满足大熊猫的进食需求。所以，它们做出了最佳的进化选择：从以吃肉为主变成了以吃竹子为主，食性发生了巨大变化。

看到大熊猫啃竹子啃得不亦乐乎，很多人认为大熊猫是植食动物，或者是杂食动物，其实不然，大熊猫在动物分类中仍然被归类为食肉目动物。早期人们将食性作为动物分类的重要参考指标，但是仅凭食性并不能对动物进行准确分类。实际上，从消化道解剖结构、生理特点以及物种进化的角度来看，大熊猫是地地道道的食肉目动物。它们是由有胎盘的肉食性哺乳动物经数百万年进化而来的。

这些进化都是为了吃竹子
Evolved to Eat Bamboo

为了更好地适应以竹子为主食的生活，大熊猫们展现出十足的生存智慧：它们增大了食量，生理结构和其功能也逐步进化，所有这些改变都是为了确保能够从竹子这种低能量食物中获取足够的营养，在地球上坚强地活下去。

To better adapt to life with bamboo as their staple food, giant pandas have shown profound survival wisdom. They have increased the amount of food intake, with their physiological structures and functions having accordingly evolved. All of those changes ensure that they are able to obtain enough nutrients from bamboo which is low in energy, to stay strong enough to survive on this planet.

伪拇指的诞生

你仔细观察过吗？大熊猫有六根"手指"。不过，严格地说，这根"第六指"并不是真正的"手指"，因为它没有"指甲"，只是一个能够活动的凸起，被称为伪拇指。伪拇指能起到类似人类大拇指的作用，与其他五指配合，让大熊猫灵活地抓握竹子。

Appearance of the Pseudo-thumb

Have you ever seen that giant pandas have 6 "fingers"? However, strictly speaking, this "sixth finger" is not really a finger because it does not have a fingernail, but a movable bulge. This is known as the pseudo-thumb, and it functions the same as the human thumb. Along with the other 5 fingers, the giant panda can grip bamboo with remarkable dexterity.

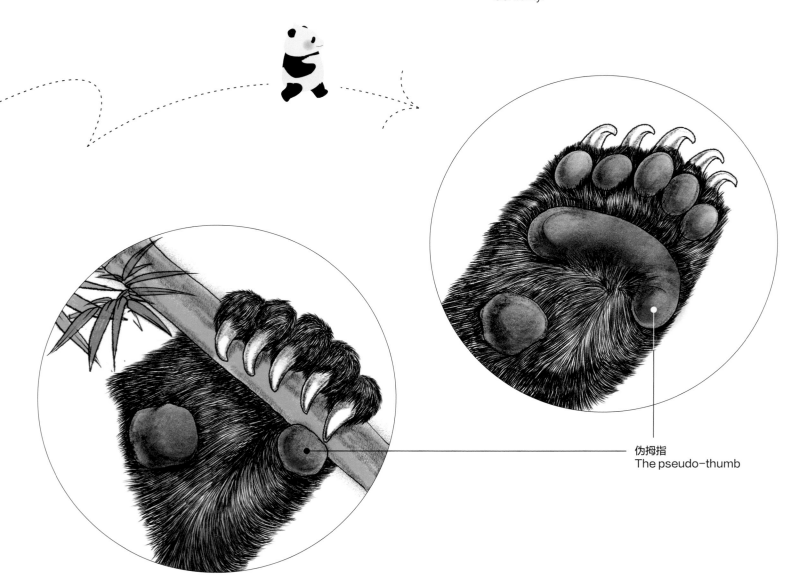

伪拇指
The pseudo-thumb

Adjustment of Dental Functions

Considerable changes have also taken place in their teeth. The canines that their ancestors used to pierce their prey are still present, but they have been blunted due to infrequent use. As their incisors have no cutting power, together with their upper cleft lips, they serve to peel off the outer skin of the bamboo poles and comb through their fur. Their large and strong molars with many tiny cusps on the surfaces facilitate crushing and masticating bamboo stems. Their cleft teeth, which were used by their ancestors for cutting tendons and tearing muscle, were also transformed into molars, with a wider chewing surface and stronger roots.

牙齿功能调整

大熊猫的牙齿也有惊人的变化。它们依然保留着祖先们用于刺杀猎物的犬齿，但是因为很少使用，犬齿已经钝化。它们的门齿已经丧失了切割的功能，主要用途是配合上裂唇，起到剥去竹竿的外皮和梳理毛发的作用。臼齿变得宽大粗壮，表面有许多细小的齿尖，有利于咀嚼研磨竹子。祖先们用于切割肉食筋腱和撕裂肌肉的裂齿，也转变为臼齿，咀嚼面增宽加大，齿根变得强壮有力。

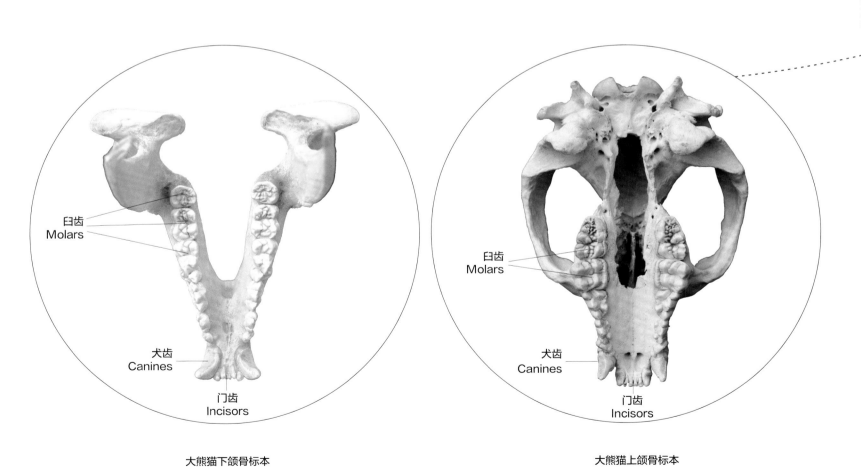

大熊猫下颌骨标本
Giant panda mandible specimen

大熊猫上颌骨标本
Giant panda maxilla specimen

颅骨全形改变

采食富含粗纤维的坚硬食物，需要有强大的咀嚼力。为了增加咀嚼肌的附着面，大熊猫的颅骨全形发生了改变，如颧弓、矢状脊等显著发达，颌关节比一般食肉动物强。

Altered Cranial Allometry

Eating hard food rich in crude fiber requires a strong chewing force. To increase the attachment surface of the chewing muscles, the cranial allometry of giant pandas has changed, with the zygomatic arch, and sagittal ridge having been significantly developed. The jaw joints are also stronger than those of the average carnivore.

大熊猫和黑熊头骨的对比
Comparison of giant panda and black bear skulls

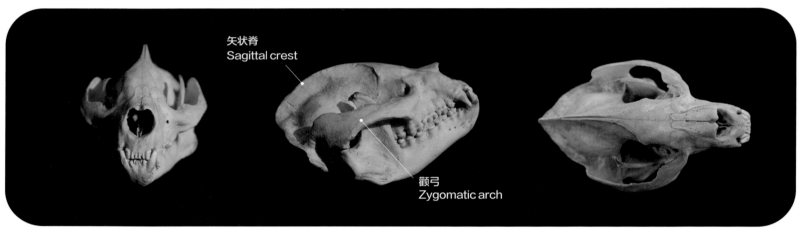

矢状脊 Sagittal crest
颧弓 Zygomatic arch

大熊猫的头骨
Giant panda skull

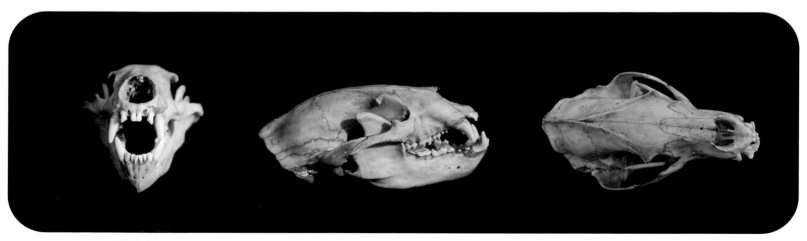

黑熊的头骨
Black bear skull

Characteristics of the Digestive System

The digestive tract of giant pandas is relatively short, unlike herbivores that have specialized chambers for storing food. They excrete feces 8 to 9 hours after eating bamboo. Although the intestinal tract is short, the digestive glands in their intestines are well developed and can secrete a large amount of mucus, which not only protects the digestive tract from being scratched by the coarse bamboo, but also acts as a lubricant. As a result, the rough bamboo residue can be glued together into a mass and be discharged without a problem, which is why giant panda feces are wrapped in a layer of mucus, conducive to excreting feces.

消化系统的特点

大熊猫的消化道比较短，不像植食动物那样有专门储存食物的腔室。它们采食竹子后 8~9 小时就会排出粪便。虽然肠道短，但它们肠道的消化腺发达，能分泌出大量黏液，一方面可以保护消化道不被粗糙的竹子刮伤，另一方面能起到润滑剂的作用，让粗糙的竹子残渣黏合成团，顺利排出。所以，大熊猫粪便外面都包裹了一层黏液，这有利于粪便的排出。

大熊猫和鹿消化系统的对比

Comparison of the digestive systems of giant pandas and deer

植食动物的肠道非常长，盲肠发达，如鹿的肠道大约是其体长的 25 倍。

Herbivores have very long intestines and well-developed cecums. For example, deer intestines are about 25 times longer than their body length.

大熊猫保留了肉食动物特有的消化道结构：肠道直且短，没有盲肠，肠道大约是其体长的 4 倍。

Giant pandas retain the characteristic digestive structure of carnivores: a straight, short intestine without a cecum, and an intestinal tract that is about 4 times their body length.

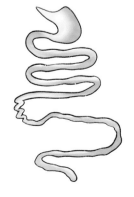

讲究的食竹攻略
Discerning Strategies for Eating Bamboo

俗话说，民以食为先。在大熊猫的世界里也是这样的。为了吃上竹子、吃好竹子，大熊猫们可是有一整套攻略的。

大熊猫 99% 的食物都是竹子。为什么是竹子？中国是全球竹类分布最丰富的地区之一，约有 40 属，700 多种。在大熊猫的野外栖息地，可供大熊猫食用的竹类植物有 12 属 60 多种。竹子分布广泛、生长快、产量高，营养成分含量虽然不高，但是一年四季常青，大熊猫很容易获取。并且，与大熊猫争食竹子的动物很少。能把自己熬成"活化石"，大熊猫这份竹子食谱真是功不可没。

有研究证明，大熊猫并不能从纤维素中获取必需的能量，而主要通过竹子中的淀粉、半纤维素等获取能量。相对其他木本植物，竹子的淀粉含量相对较高，而且竹子不同部位的淀粉含量随季节的变化也会发生变化，大熊猫总是选择取食竹子中淀粉含量高的部位。

竹子中淀粉和半纤维素含量最高的是竹笋。每年的发笋季节，也是大熊猫发情和生儿育女的季节。在这个季节里，大熊猫优先选择取食竹笋。而在没有竹笋和嫩叶的情况下，竹竿中的淀粉含量及可溶性糖的含量则会达到一年中的最高点，这时大熊猫就开始食用竹竿。

As the Chinese saying goes, "It's imperative to eat your fill." The same is true of giant pandas. To eat bamboo, especially fine species, pandas have a complete set of strategies.

Giant pandas almost entirely feed on bamboo. Why bamboo? China is one of the richest regions in the world in terms of bamboo distribution, with more than 700 species across about 40 genera. In the giant panda habitats, there are more than 60 kinds of bamboo plants of 12 genera that are edible. Bamboo is widely distributed, fast-growing, and high yielding. Although it is low in nutrients, it is evergreen throughout the year, which makes it easily accessible for pandas. Moreover, there are few animals that compete with pandas for bamboo. As a species that has survived for millions of years, pandas attribute their existence partly to bamboo recipe.

Some studies have proved that pandas cannot get the necessary energy from cellulose, but mainly through the bamboo starch, and hemicellulose. Compared with other woody plants, the starch content of bamboo is relatively higher, and that of different bamboo parts varies with the seasons. Giant pandas always choose to eat the parts with a high starch content.

Bamboo shoots contain the highest starch and hemicellulose content. Every year, the bamboo shoot season coincides with the season that giant pandas are in heat and have cubs. During this season, giant pandas prefer to feed on bamboo shoots. In the absence of bamboo shoots and tender leaves, the starch content and soluble sugar content in bamboo stems will reach its peak for the year, and then the giant pandas will turn to bamboo stems.

吃货们,"赶笋"去啦!
>> Foodies Catching Bamboo Shoots

大熊猫喜食的竹类随山系和季节不同而有所变化。它们优先采食营养最优的竹子种类或竹子部位,最大限度地从食物中获取营养和能量,并通过季节性的食物转换来满足营养和能量需求以达到营养均衡。

新鲜脆嫩、口感上佳的竹子和竹笋最受大熊猫青睐。它们会根据季节的变化,从低谷到高山垂直迁徙采食新鲜竹子和竹笋,这被形象地称为"赶笋"。

The types of bamboo that giant pandas prefer to eat vary according to the mountain system and the seasons. They prioritize the most nutritious bamboo species or parts of the bamboo, maximizing the nutrition and energy from it. Apart from that, they keep a well-balanced diet by changing their food according to the different seasons to meet their needs of energy and nutrition.

Fresh, crispy, and tasty bamboo and bamboo shoots are their favorite. They will migrate vertically from low valleys to high mountains to forage for fresh bamboo and bamboo shoots based on seasonal changes, a phenomenon which is figuratively called "catching bamboo shoots".

多吃少动只为节能
Saving Energy by Consuming More and Moving Less

有人看大熊猫一天到晚都在吃，吃完就睡觉，觉得它们好吃懒做。这真是天大的误会呀！大熊猫每天花10小时或更多时间来进食，但是你知道吗，根据所食用的竹叶、竹竿和竹笋的不同，一只成年大熊猫每天要采食10~40千克鲜竹，并且在8~9小时后就排出粪便。因为大熊猫以低营养和低能量的竹子为主食，它们的能量代谢率异常低，还不到同体重哺乳动物的一半，所以它们只有选择多吃少动，才能维持庞大身体的能量需要。看似慵懒的大熊猫，其实是动物界妥妥的节能标兵。

Pandas are seen eating almost all day long and sleeping after eating their fill, making people think they are lazy. This is a complete misunderstanding! Giant pandas spend 10 hours or more a day eating, but what is less known is that an adult giant panda will consume 10 to 40 kilograms of fresh bamboo per day according to different types of bamboo leaves, stems, and shoots, and excrete feces in 8 to 9 hours. Because pandas have a bamboo diet which is low in nutrients and energy, their metabolic rate is abnormally low, less than half that of a mammal of the same weight. Thus, they have no choice but to eat more and move less to meet the energy needs of their large bodies. While seemingly lazy, in fact, it is a vivid example of energy-conservation in the animal world.

进食
Eating

8 ~ 9 小时后
8 to 9 hours later

排便
Defecation

大熊猫的便便是"香"的
>> "Fragrant" Poop

大熊猫的便便呈梭形，根据所吃的竹叶、竹竿、竹笋的不同，便便也会呈现不同的颜色。吃竹叶的便便呈绿色，吃竹竿的便便呈黄绿色，吃竹笋的便便呈淡黄色。

因为大熊猫的消化道短，竹子在肠道里不会被完全消化吸收，也来不及发酵，所以它们的便便呈现出的是未经完全消化的竹子残渣的样子，不仅不臭，还有一股淡淡的竹子清香。

研究人员可以根据粪便的颜色和其中残留的竹节长短来判断大熊猫的年龄和健康状况，还会通过研究粪便的成分，探析大熊猫肠道内的菌群组成。

成年大熊猫吃竹笋时一天的排便量在20～40千克，吃竹竿和竹叶时一天的排便量为10～20千克。它们的便便可回收加工成纸张或工艺品出售，还可以作为肥料用来种植农作物。

Giant panda stools are shuttle-shaped and vary in color based on the types of bamboo leaves, bamboo stems, and bamboo shoots they eat. Those stools of pandas eating bamboo leaves are green, bamboo stems yellowish green, and bamboo shoots light yellow.

Because of their short digestive tract, the intestine can neither completely digest and absorb bamboo, nor ferment bamboo, which explains why their poop looks like semi-digested bamboo residue. However, it does not stink, and there is a light fragrance of bamboo instead.

Researchers can tell the age and health conditions of pandas based on the color of the feces and the length of the remaining bamboo parts. Besides, they can study the composition of the feces and analyze the flora in the pandas' intestines.

Adult giant pandas defecate 20 to 40 kilograms a day when they eat bamboo shoots, while 10 to 20 kilograms a day when they eat bamboo stems and leaves. The waste can be recycled and processed into paper or crafts for sale, and can also be used as fertilizer to nurture crops.

吃竹笋的便便 / Feces from eating bamboo shoots
吃竹竿的便便 / Feces from eating bamboo stems
吃竹叶的便便 / Feces from eating bamboo leaves

大熊猫的食谱：野生和圈养的不一样
Recipes for Giant Pandas: Different in the Wild and in Captivity

虽然大熊猫以竹子为主食，但在野外它们偶尔也会食用一些其他植物的枝叶、果实，以及一些动物的尸体组织或骨骼等，通过对食物进行充分选择来实现营养的均衡。

Although giant pandas mainly feed on bamboo, in the wild they occasionally consume branches, leaves, and fruits of other plants, as well as the tissues or bones from carcasses of some animals, to have a balanced intake of nutrition by diversifying their choices of food.

成年野生大熊猫的食谱
Recipes for adult wild giant pandas

- 竹叶 bamboo leaves
- 竹竿 bamboo stems
- 竹笋 bamboo shoots
- 树叶 leaves
- 野果 wild fruits
- 动物尸体组织或骨骼 tissues or bones from carcasses

99% 的食物为竹子
99% is bamboo

成年圈养大熊猫的食谱
Recipes for adult captive giant pandas

- 竹叶 bamboo leaves
- 竹竿 bamboo stems
- 竹笋 bamboo shoots
- 大熊猫窝窝头 panda cake (wowotou)
- 苹果 apples

99% 的食物为竹子
99% is bamboo

大熊猫窝窝头是以玉米、大米、燕麦、大豆、小麦为主，添加植物油、微量元素和维生素等蒸制而成的大熊猫辅食。

Panda cake (wowotou) is a complementary food for giant pandas steamed with corn, rice, oats, soybeans and wheat, supplemented with vegetable oil, trace elements and vitamins.

Breeding and Growth of the Giant Panda
大熊猫的孕育和成长

圈养条件下，大熊猫的平均寿命为 20~25 岁，相比人类，这样的"熊生"不算长。但是，从粉嘟嘟的迷你宝宝，到换上标志性的"黑白外套"逐步成长，追逐玩闹、谈情说爱、孕育后代，大熊猫的这一生，同样充满了酸甜苦辣、喜怒哀乐，同样丰富多彩。

Captive giant pandas enjoy an average life expectancy of 20 to 25 years, which is far shorter compared to human beings. However, their colorful life is also full of sorrows and joys as well, like growing from the pink tiny cub to the one with its iconic "black and white coat", chasing and playing, as well as mating and breeding.

Breeding: Once a Year

Every year in March to May, when flowers bloom, it is the season for giant panda estrus. Giant pandas are in heat only once a year, and each time it can last for 10 to 20 days, during which only 1 to 3 days can witness a successful conception. Harder still, the optimal time for conception is only a few hours. If they miss that time, they will have to wait another year.

Courtship
During estrus, solitary female giant pandas leave scents in their environment and emit vocal signals to attract mates. Males go through a certain degree of fighting to gain the right to mate with females.

Mating
Mating in giant pandas usually takes place in the mountains and sometimes in trees. After the process, the males and females will return to their solitary lifestyles.

Pregnancy
Generally the gestation period lasts about 5 months, with the shortest being 68 days and the longest 324 days. During this period, they need to eat a lot to meet the nutritional needs of both themselves and the fetus. Several days before giving birth, they basically cease eating or drinking.

每年的 3~5 月，春暖花开，正是大熊猫们发情的季节。大熊猫的发情期一年只有一次，每次可以持续 10~20 天，但是只有 1~3 天可以受孕，最佳受孕时间仅为几小时。如果错过了这个时间，那就要再等上 1 年了。

求偶
发情期间，独居的雌性大熊猫在环境中留下气味标记信息，发出声音信号，用于寻找配偶。雄性个体需通过一定程度的争斗，才能获得与雌性的交配权。

交配
大熊猫的交配通常在山地原野进行，有时也在树上进行。交配完成后，雌雄个体各自回归独居的生活方式。

孕期
大熊猫的孕期一般为 5 个月左右，最短的只有 68 天，最长的达到过 324 天。在怀孕期间，它们需要大量进食，以满足自身和胎儿的营养需要，到了临产的前几天，则基本不吃不喝。

雄性大熊猫通过争斗取得与雌性的交配权
Male pandas fight for the right to mate with females

产仔：大约在秋季
>> Giving Birth: In Autumn

大熊猫一般在秋季产仔。临产前，雌性大熊猫一般会选择树洞或者岩洞作为巢穴，在里面铺垫好树枝或干草，让巢穴变得温暖舒适。不管选择的是树洞还是岩洞，周围的环境都一定要比较安静，竹林茂密，隐蔽条件较好，离水源也比较近。当生产条件都准备好后，雌性大熊猫的活动开始减少，静静等待熊猫宝宝的出生。

幼仔出生后，如果是双胞胎，为了保证在野外艰苦的条件下至少能存活一只，大熊猫妈妈一般会选择身体强壮的一只幼仔来抚养，另一只则会被抛弃。

Giant pandas usually give birth in the autumn. Before giving birth, female giant pandas will usually choose a tree den or a rock cave as their nest, where they will lay down branches or grass to make the nest warm and cozy. Regardless of their choices, the surrounding environment must be quiet and close to water sources, with dense bamboo forests for concealment. When they have prepared for birth, they will become less active and quietly wait for the delivery.

After the cubs are born, if they are twins, to ensure that at least one survives in the harsh conditions of the wild, the mother will usually choose the physically stronger one to raise, while abandoning the other.

圈养条件下，饲养员会采用"换仔技术"辅助大熊猫妈妈照顾双胞胎宝宝。

In captivity, keepers will use "baby-switching technique" to assist the mother in caring for her twins.

育幼：妈妈寸步不离
>> Nurturing: Incessant Care

大熊猫的初生幼仔很小，平均体重为 120 克左右，约为熊猫妈妈体重的千分之一。才出生的大熊猫幼仔是粉红色的，身上长有稀疏的白毛，因为毛发稀少，身体发育又差，无法自主维持体温，所以大熊猫妈妈会一直把幼仔抱在怀里温暖它。

新生幼仔的消化系统发育不完全，所以无法进行自主排便，大熊猫妈妈会用舌头不断舔舐幼仔的会阴部位辅助它排便。因为大熊猫幼仔出生时发育不完全，是地地道道的早产儿，很容易因为体温下降、缺乏营养、遭遇敌害等原因夭折，所以大熊猫妈妈在幼仔出生后几天内都不吃不喝，寸步不离地照顾幼仔。熊猫幼仔出生后趴在大熊猫妈妈的腹部吮吸乳汁，吃完一边又换另一边，大熊猫妈妈的初乳为新生幼仔提供了必需的营养和免疫抗体。

在大熊猫妈妈的照顾下，大熊猫宝宝快速成长，逐步学会爬行、走路、吃竹子、爬树、交流等各种本领。

Giant panda cubs are very small, weighing about 120 grams on average, which is about one-thousandth of the mother's weight. The newborn giant panda cub is pink with sparse white hair. Due to the scarcity of hair and the poor development of their body, they cannot maintain their own body temperature, so the mother will always hold the cub in her arms to warm them.

The newborn cub does not have a fully-developed digestive system, which makes it unable to defecate on its own, so the mother will continually lick the perineum of the cub to assist it in defecating. Rudimentary development caused by premature birth leaves it prone to death due to drops in body temperature, a lack of nutrients, and exposure to threats, so the mother will not eat or drink for a few days after giving birth only to care for the cub incessantly. The newly born cub usually lies on the mother's belly and sucks on the colostrum, and after finishing on one side, it switches to the other, from which it gains essential nutrients and antibodies.

Under care of the mother, the cub grows rapidly and gradually learns various skills such as crawling, walking, eating bamboo, climbing trees, and communicating.

大熊猫初乳呈现淡淡的绿色
Giant panda colostrum is light green.

世界上初生体重最轻并存活的大熊猫是 2019 年 6 月出生在成都大熊猫繁育研究基地的"成浪"，它出生时仅重 42.8 克。

The world's lightest new-born panda cub who survives is Cheng Lang, born in June 2019 at the Chengdu Research Base of Giant Panda Breeding, weighing only 42.8 grams.

成长：约 1 岁半离开妈妈
Growth: Leaving Mom at about One Year and a Half

大熊猫在 1.5 岁～2 岁时断奶，然后就会离开妈妈独自生活。

断奶后到性成熟前为大熊猫的亚成年阶段，相当于人类的青少年时期。这时的大熊猫生理和心理上逐渐成熟。大约在 5～6 岁时，它们步入成年期，开始谈情说爱、繁育后代。

Giant pandas are weaned at the age of 1.5 to 2, and then leave their mothers to start their own lives.

The period after weaning to before sexual maturity is their sub-adulthood, which likes adolescence in human beings. At that stage, giant pandas gradually become mature both physically and mentally. At around 5 to 6 years old when they enter adulthood, they begin to mate and reproduce.

大熊猫成长日记
Growth Diary of Giant Pandas

120 g

刚出生
Newborn

刚出生时，我的皮肤是粉色的，并带有稀疏白毛，体重仅是妈妈的千分之一。

At birth, I am pink with sparse white fur and weigh only a fraction of my mom.

我的眼眶、耳朵、肩带和四肢的颜色开始慢慢变深。

The color of my eye sockets, ears, shoulders, and limbs began to darken slowly.

15 天左右
Around 15 days

600 g

我睁开眼，终于看到了外面的世界。
I open my eyes and finally see the outside world.

50 天左右
Around 50 days

2.2 kg

1.2 kg

30 天左右
Around 30 days

我变得黑白分明，更像妈妈了。
Iconic black-and-white markings begin to emerge, and I look more like my mom.

4 个月左右
Around 4 months

7 kg

我开始爬动、学习走路，妈妈也会和我玩耍，帮助我锻炼行为能力。
I start crawling and learning to walk, and my mom plays with me to help me develop.

第二章 / 竹林隐士

6 个月左右
Around 6 months

10 kg

我的食物还是以奶为主，但我已经开始向妈妈学习吃竹子。我还喜欢上了爬树。

I still mainly feed on milk, but I've started learning to eat bamboo from my mom. Besides, I also enjoy climbing trees.

我会和妈妈一起生活到一岁半至两岁，然后我就要离开妈妈，开始独立生活了。

I will live with my mom until I am 1.5 to 2 years old, and then leave her to start living on my own.

1.5 ~ 2 岁
Around 1.5 to 2 years old

50 kg

5 ~ 6 岁
Around 5 to 6 years old

80 ~ 120 kg

我终于成年了，有了自己独居的领地，开始寻觅我的"另一半"。

I finally enter adulthood and have my own territory, and then I will begin looking for my own mate.

熊猫世界

大熊猫的寿命和年龄划分
>> Lifespan and Age Classification of Giant Pandas

初生阶段
Newborn Stage

在野外，大熊猫的平均寿命为 18~20 岁。圈养条件下，大熊猫的平均寿命为 20~25 岁，有的甚至超过 30 岁。

大熊猫 1.5 岁前为幼年期，雌性大熊猫一般 5 岁左右成年，雄性大熊猫一般 6 岁左右成年，幼年期和成年期之间为亚成年期。

幼年阶段
Infant Stage

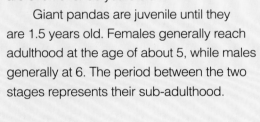

In the wild, the average lifespan of a giant panda is 18 to 20 years old. In contrast, the average lifespan of a captive giant panda is 20 to 25 years old, and some are even over 30 years old.

Giant pandas are juvenile until they are 1.5 years old. Females generally reach adulthood at the age of about 5, while males generally at 6. The period between the two stages represents their sub-adulthood.

亚成年阶段
Sub-adult Stage

成年阶段
Adult Stage

Giant Panda Habits

大熊猫的生活习性

I am a Recluse

Except for a few days when male and female pandas briefly get together during the mating season, adult giant pandas live alone almost all year round, a typical solitary animal.

Excellent Tree Climber

With sharp claws that can easily dig into tree bark as well as the fur and pads around them that stop slipping, giant pandas are able to firmly grasp the trunks. Plus, their developed and powerful limbs enable them to climb trees quickly. Climbing trees is an essential skill for them during their juvenile years. When their mother goes out to forage for food, panda cubs will climb high up in the trees to hide from their natural enemies.

Wherever I am Is Home

Wild giant pandas have no fixed resting place, sometimes in tree dens and rock caves and under large trees or in bamboo forests. In the wild, pregnant giant pandas prefer covert tree dens or rock caves and carefully lay dead leaves and hay inside. They will also live in those places for months after giving birth.

Hibernation? Not in My Dictionary!

Unlike brown bears and black bears, giant pandas do not hibernate, and even in harsh winters, they forage in the snow as usual. This is partly because their fur is warm and moisture-proof, allowing them to adapt well to the cold and wet natural environment. Apart from that, plenty of bamboo throughout the year does not necessitate the need to hibernate.

我是独行侠

除了在发情交配季节雌雄大熊猫会短暂相聚几天外，成年大熊猫几乎常年独居，是典型的独居动物。

爬树是高手

大熊猫有着锋利的爪子，爪尖可以轻易地嵌入树皮；同时爪子周围的毛发、肉垫能够防滑，从而帮助它们牢牢地抓抱树干；发达有力的四肢使它们能够快速爬上大树。爬树是大熊猫幼年时期一项非常重要的技能。当大熊猫妈妈外出觅食，大熊猫宝宝就会爬到高高的树上躲避天敌。

我在哪儿，家就在哪儿

野外大熊猫无固定休息场所，有时在树洞、岩洞里休息，有时在大树下或竹林中休息。在野外，怀孕的大熊猫妈妈会选择隐蔽的树洞或岩洞，在里面精心铺垫枯叶和干草，在宝宝出生后的几个月中在此栖身。

冬眠？不需要！

大熊猫不像棕熊、黑熊有冬眠习性，即使在严冬，它们也照常在雪地中觅食。这一方面是因为它们的皮毛保暖又防潮，让它们能很好地适应严寒、潮湿的自然环境；另一方面是因为一年四季都有竹子可以采食，食物充足，当然就不需要冬眠了。

Giant Panda Family and Genus
大熊猫的科属

大熊猫属于什么科?
What Family Do Giant Pandas Belong to?

Scientists have not reached a consensus about the classification of giant pandas. As for red pandas, which were once considered to be a part of the raccoon family and share the giant pandas' characteristic pseudo-thumb, some experts believe that the giant panda should also belong to the raccoon family. However, red pandas were later placed into the independent family Ailuridae. By the same token, according to the natural history and physiological characteristics of the giant panda, researchers also put the giant panda into the separate family Ailuropodidae. With the development of molecular biology, a growing body of evidence suggests that the giant panda is closer to the bear and should belong to the family Ursidae.

关于大熊猫的分类,科学家们一直争论不休。小熊猫(红熊猫)曾被认为属于浣熊科,而大熊猫特有的伪拇指结构与小熊猫类似,据此,一些专家认为大熊猫也应属于浣熊科。后来,小熊猫被独立为了小熊猫科,而根据大熊猫的自然历史、生理特性等,研究者们也将大熊猫独立为了大熊猫科。现在,随着分子生物学的发展,更多证据表明大熊猫与熊更为接近,应属于熊科。

熊科动物的共同特点
Common Characteristics of Animals in the Family Ursidae

They are usually stout and large with a round head, small eye sockets, short neck, short tail, strong limbs with 5-toes on the front and rear paws as well as sharp claws. Existing animals in the family Ursidae have deviated from their meat-only diet nature, and most have evolved into omnivores, which has blunted their teeth. However, the crowns of their molars are flat and wide - a structure suitable for grinding food.

它们通常体形肥壮,头大而圆,眼眶小,脖子短,尾巴短,四肢粗壮,前后足都有5趾,爪子锋利。现生的熊科动物已经偏离了只吃肉的特性,多数种类已经特化成为杂食动物,所以它们的牙齿不尖锐,但是臼齿齿冠平而宽——这种结构适用于研磨食物。

熊科包含：

3 个亚科
5 个属
8 个物种

The Family Ursidae Consists of:

3 subfamilies
5 genera
8 species

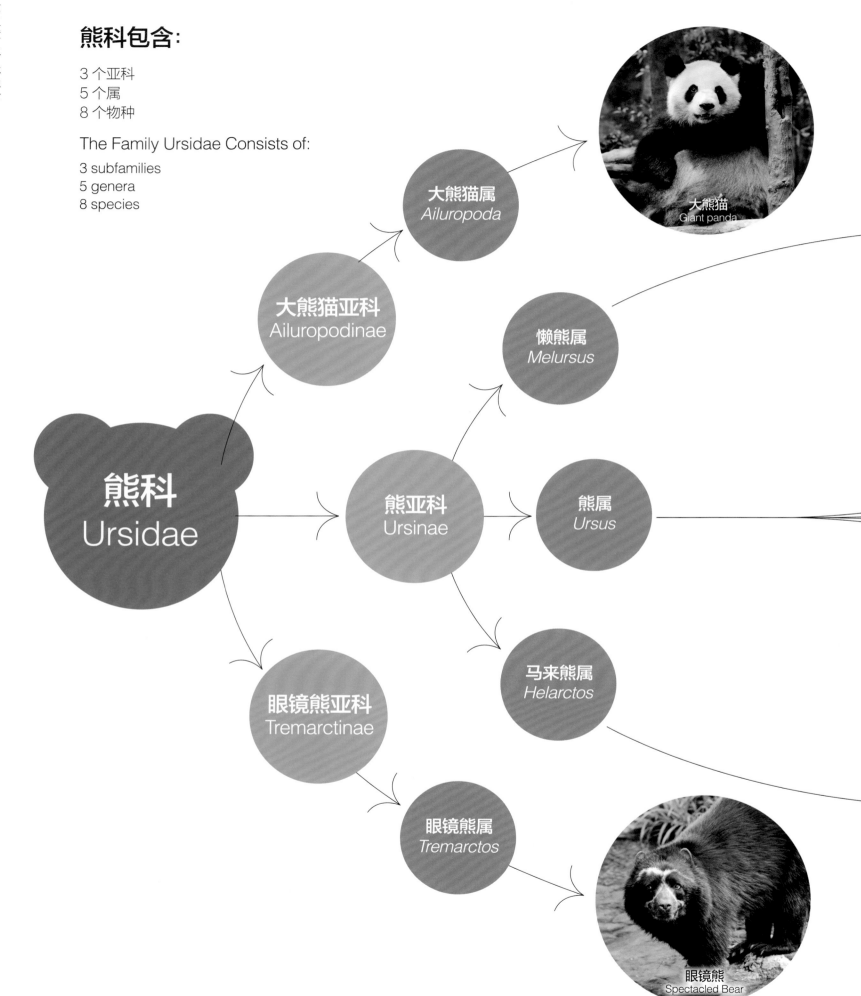

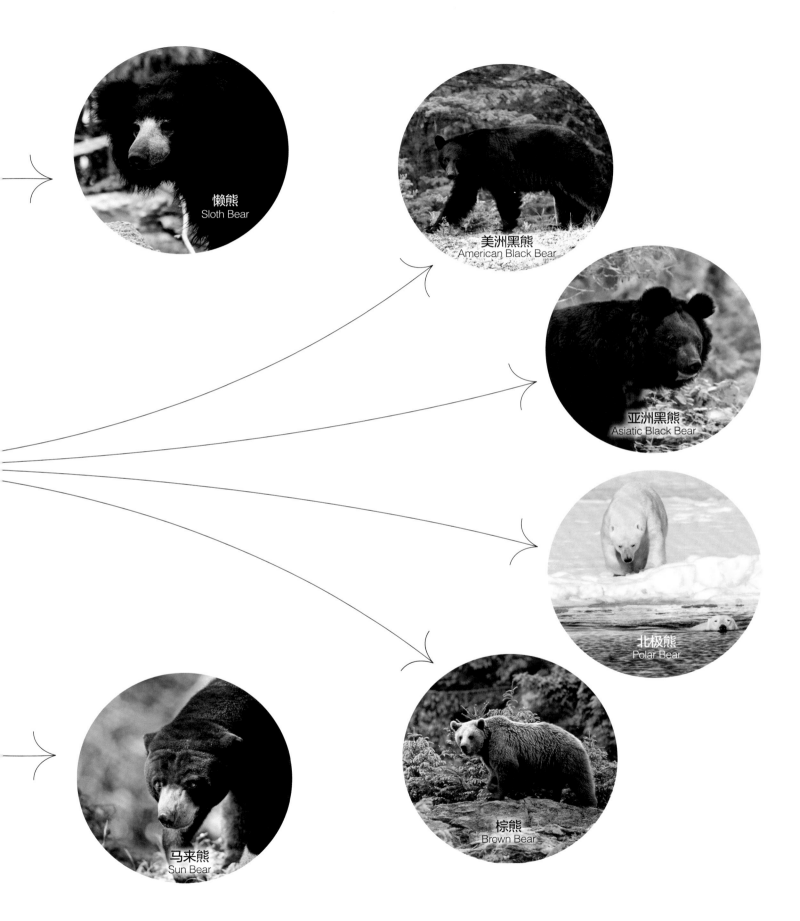

眼镜熊
Spectacled Bear
(*Tremarctos ornatus*)

Spectacled Bear
I have dark circles under my eyes too

Food: Omnivorous. With a primarily plant-based diet, they are still powerful hunters, capable of preying on large animals such as deer, llamas, domestic cattle, and horses; and small ones such as rabbits, mice, as well as birds. They also sometimes eat carrion.
Characteristics: *Tremarctos ornatus* is covered with blackish fur with pale markings across its face and fore chest and a pair of white rings around the eyes like glasses. They have a keen sense of smell, moderate hearing but weak eyesight. They excel at climbing and often build platforms in the canopy for resting, avoiding danger, and storing food. They are solitary animals, but have no sense of territory. They do not hibernate because they have sufficient food year-round.
Distribution: *Tremarctos ornatus*, also known as the Andean bear, is endemic to and the only family Ursidae animal in South America. They are usually found in Venezuela, Colombia, Ecuador, Peru, and Bolivia.
Lifespan: 20 to 25 years in captivity.
Protection class: Vulnerable Species on the IUCN Red List.

眼镜熊
我也有黑眼圈

食性： 杂食，以植食为主，但仍是强大的猎手，能捕食鹿、美洲驼、家牛、马等大型动物和兔、鼠等小型动物以及鸟类，有时也吃腐肉。
特点： 眼镜熊具有黑色的体毛，脸部和前胸部为白色，眼睛周围有一对像眼镜一样的白圈。眼镜熊嗅觉灵敏，听力中等，视力较弱。它们具有高超的攀爬能力，常在树冠上搭建平台，用来休息、躲避危险、储存食物。眼镜熊是独居动物，但没有领地意识。因为常年食物充足，所以它们不冬眠。
分布： 眼镜熊也叫安第斯熊，是南美洲特有的熊科动物，也是南美洲唯一一种熊科动物，分布于委内瑞拉、哥伦比亚、厄瓜多尔、秘鲁和玻利维亚。
寿命： 圈养状态下一般寿命为 20~25 年。
保护等级： 世界自然保护联盟（IUCN）红色名录易危物种。

懒熊
我拿蚂蚁当主食

食性： 杂食，以白蚁等昆虫为主食，也吃树叶、花朵、水果、谷物和小型脊椎动物。

特点： 懒熊全身覆盖着长长的黑毛，前胸点缀着一块白色或淡黄色的"U"形或者"Y"形斑纹。它们没有中央门齿，嘴唇灵活而且可以伸缩，鼻孔在吸食昆虫时可以封闭，前肢有大而弯曲的爪，善于扒开蜜蜂和白蚁等昆虫的巢。懒熊是一种独居动物，不冬眠，主要在夜间活动。

分布： 在中国分布于西藏藏南地区，国外主要分布于印度、斯里兰卡的森林中，在孟加拉国、尼泊尔和不丹也有少量分布。

寿命： 人工饲养条件下寿命可达 40 年。

保护等级： 世界自然保护联盟（IUCN）红色名录易危物种，中国国家二级重点保护野生动物。

Sloth Bear
I Take Termites as My Staple Food

Food: Omnivorous. They mainly feed on insects like termites, but also eat leaves, flowers, fruits, grains, and small vertebrates.

Characteristics: The Sloth Bear is covered with long black fur and has a white or yellowish "U" or "Y" shaped markings on the fore chest. They lack central incisors, but their lips are flexible and retractable. Their nostrils are closed when sucking insects. They have large and sickle-shaped claws, which enable them to pick open the nests of insects such as bees and termites. The Sloth Bear is a solitary animal, does not hibernate, and is nocturnal.

Distribution: They are found in the southern part of Xizang, China. The Sloth Bear's global range includes the forests of India and Sri Lanka, and a small amount of them are also found in Bangladesh, Nepal, and Bhutan.

Lifespan: Up to 40 years in captivity.

Protection class: Vulnerable Species on the IUCN Red List and the wildlife under state second-class key protection in China.

马来熊
Sun Bear
(Helarctos malayanus)

Sun Bear
I'm the Smallest Bear Species with the Shortest Fur

Food: Omnivorous. They feed on a broad variety of things such as termites, ants, beetle larvae, bee larvae (including honey), invertebrates, and various types of plant fruits. They occasionally prey on small vertebrates such as reptiles and birds, as well as bird eggs.

Characteristics: Also known as sun bears, they are the smallest in the family Ursidae. Their fur is generally jet black, with an orange or gray patch around the eyes to the muzzle, and a large white to yellowish patch on the chest, typically U-shaped, which is their characteristic markings. With proportionately smaller ears, sun bears have elongated limbs and long, powerful claws, which equip them with the ability to climb trees well.

Distribution: They are mainly found in Indonesia, the Malay Peninsula, and Myanmar. There are only a few in southwestern Yunnan, China.

Lifespan: About 20 years.

Protection class: Vulnerable Species on the IUCN Red List, and the wildlife under state first-class key protection in China.

马来熊
我个头最小，毛毛也最短

食性： 杂食，食物包括白蚁、蚂蚁、甲虫幼虫、蜜蜂幼虫（包括蜂蜜）等，以及无脊椎动物和各类植物果实，偶尔捕食爬行类、鸟类等小型脊椎动物，也吃鸟卵。

特点： 又叫太阳熊，是体形最小的熊科动物。整体毛色为黑色，眼周围至吻鼻部呈橙黄色或灰色，双耳非常小，胸部有一块大型的白色至乳黄色块斑，通常为"U"形，可作为个体识别的标志。马来熊四肢修长，爪长而有力，具有极强的爬树能力。

分布： 它们主要分布在印度尼西亚、马来半岛、缅甸等地。在中国仅有云南西南部有少量分布。

寿命： 寿命一般为 20 年左右。

保护等级： 世界自然保护联盟（IUCN）红色名录易危物种，中国国家一级重点保护野生动物。

美洲黑熊
American Black Bear
(Ursus americanus)

美洲黑熊
遇危险，我唰唰蹿上树

食性：杂食，以植食为主，主要吃禾本科植物、草本植物、水果，以及菌类、植物的果实、种子等，也吃昆虫、鱼、蛙、鸟卵及小型兽类，喜欢挖蚂蚁窝和掏蜂巢。

特点：美洲黑熊四肢粗短，长着细小的眼睛、圆耳朵、长鼻子和短尾巴。多数美洲黑熊的体毛为黑色，但有的亚种有不同的毛色，比如白色、棕色、金黄色。还有些美洲黑熊的胸部长有白纹。它们的掌上长有五根强壮的爪，主要用来撕扯、挖掘和攀爬。美洲黑熊喜欢生活在森林和灌丛中，当它们遇到危险时会爬到树上，利用森林地区作为行走回廊。它们有冬眠的习性，并会在树洞、河堤边、山洞等地方建立巢穴。

分布：广泛分布于北美洲。

寿命：寿命一般为 25 年，人工圈养条件下一般为 35 年。

保护等级：世界自然保护联盟（IUCN）红色名录无危物种。

American Black Bear
When in Danger, I Leap up Trees

Food: Omnivorous. Their diet consists of vegetation, mainly Poaceae Barnhart, herbaceous plants, fruits, fungi, plant fruits and seeds. It also includes insects, fish, frogs, bird eggs, and other small animals. They like to hollow out ant nests and bee hives.

Characteristics: The American Black Bear has short limbs, small eyes, round ears, a long nose, and a short tail. Most have black fur, but some subspecies have different colors, such as white, brown, and blonde. There are also some of them with white patches on their chest. They have 5 sharp claws on their paws, which are mainly used for tearing, digging, and climbing. The species prefers to live in forests and thickets. When they are in danger, they will flee up trees, taking advantage of their familiarity with the area to escape threats. They are hibernators and tend to build dens in tree cavities, along riverbanks, and in caves.

Distribution: Widely seen in North America.

Lifespan: 25 years on average, and 35 years in captivity.

Protection class: Least concern Species on the IUCN Red List.

亚洲黑熊
Asiatic Black Bear
(*Ursus thibetanus*)

Asiatic Black Bear
The "Bear Blind" Is Me.

Food: Omnivorous. Herbivorous to a great degree, they feed mainly on fresh plants in the spring, plant fruits in the summer, and nuts in the fall. They also relish small animals and occasionally eat carrion.
Characteristics: They are also known as moon bears because of the crescent-shaped white chest patch on their chest. Although they have a strong sense of smell, their eyesight is poor, and their hearing range is moderate. Therefore, they are dubbed "bear blind". The vast majority of them do not hibernate, while only those living in the cold north will do so. They are good climbers, spending half their lives in trees. They can build dens in tree holes or trees.
Distribution: Often found in Asia.
Lifespan: On average 25 years in the wild.
Protection class: Vulnerable Species on the IUCN Red List, and the wildlife under state second-class key protection in China.

亚洲黑熊
"熊瞎子"就是我

食性： 杂食，以植食为主，春季主要以鲜嫩多汁的植物为食，夏季会选择植物的果实，秋季则常取食坚果。它们对小型动物也感兴趣，偶尔还吃腐肉。
特点： 亚洲黑熊因胸部有月牙形的白色胸斑，又叫月熊。它们嗅觉很强，听觉范围适中，但视力却很差，因此被人们称为"熊瞎子"。绝大多数亚洲黑熊是不冬眠的，只有生活在寒冷北方的亚洲黑熊才有冬眠的习性。亚洲黑熊善于攀爬，它们一生中有一半的时间是在树上度过的。它们能在树洞或树上筑巢。
分布： 分布于亚洲地区。
寿命： 野外平均寿命为 25 年。
保护等级： 世界自然保护联盟（IUCN）红色名录易危物种，中国国家二级重点保护野生动物。

北极熊
我的皮肤其实是黑色的

食物： 北极熊是食肉程度最高的熊类，主要食物是海豹。除此之外，它们也捕食鱼类、鸟类等。在夏季它们偶尔会吃点儿浆果或者植物的根茎。

特点： 北极熊是世界上最大的陆生食肉动物。虽然看似浑身雪白，但是北极熊的皮肤其实是黑色的，而且它们拥有一层用于保暖的绒毛和一层长达15厘米的空心透明的硬针毛，两者结合起来具有极强的吸热保暖能力，所以北极熊能在寒冷的北极地区自在地生活。北极熊体形巨大，成年雄性体重300~800千克，成年雌性体重150~450千克。冬季临近时它们还需要囤积大量脂肪。这层厚厚的脂肪给了它们更强的浮力，让它们在水中来去自如。看似笨重的北极熊，可是十分出色的游泳健将。

分布： 分布于丹麦的格陵兰岛、美国的阿拉斯加，以及挪威、加拿大、俄罗斯等国的环北极地区。

寿命： 寿命一般为30年左右

保护等级： 世界自然保护联盟（IUCN）红色名录易危物种。

Polar Bear
In Fact, My Skin Is Black

Food: The Polar Bear is the most carnivorous species of bear, with seals as their staple. In addition, they also prey on fish and birds. In the summer, they occasionally eat berries or plant roots.

Characteristics: The Polar Bear is the largest living species of land carnivore. Although it looks white, its skin is actually black. The coat consists of dense underfur for warmth and hollow transparent guard hair around 15 cm, the combination of which can trap heat and keep it warm, so polar bears can live freely in the cold Arctic region. Polar bears are huge, with adult males weighing 300 to 800 kilograms and adult females weighing 150 to 450 kilograms. They need to store large amounts of fat as winter approaches in a thick layer, which gives them greater buoyancy and allows them to move freely in the water. Although they look bulky, they are excellent swimmers.

Distribution: Greenland in Denmark, Alaska in the United States, and the circumpolar regions of Norway, Canada, Russia and other countries.

Lifespan: On average 30 years

Protection class: Vulnerable Species on the IUCN Red List.

棕熊
Brown Bear
(*Ursus arctos*)

Brown Bear
My Family Is Found Across Eurasia and North America

Food: Omnivorous. Their diet consists of various types of herbaceous plants, plant bulbs, tubers, grains, fruits, etc. They also prey on hoofed animals, and occasionally ingest insects and carrion. They also like to fish.

Characteristics: They have a hump at the top of their shoulder, which is made entirely of muscle. Their fur is variable, including gray-black, brown-black, dark brown, brown-red, yellowish-brown, and gray, while brown is the most common. Their limbs' fur is nearly black in color. Occasionally, white individuals can be seen.

Distribution: The Brown Bear is the most widely distributed species of the family Ursidae in the world, primarily found on the most parts of Eurasian continent in the Northern Hemisphere, as well as the northern and western parts of the North American continent. In China, it is mainly found in Heilongjiang, Jilin, Liaoning, Inner Mongolia, Xinjiang, Gansu, Qinghai, Xizang, and Sichuan.

Lifespan: On average 30 years in the wild.

Protection class: Least concern Species on the IUCN Red List, and the wildlife under state second-class key protection in China.

棕熊
我的家族遍布亚洲、欧洲和北美洲

食物： 杂食，其食物包括各类草本植物、植物球茎、块茎、谷物、果实等，它们也捕食有蹄类动物，偶尔摄食昆虫，并且喜欢捕鱼，有时也吃腐肉。

特点： 棕熊肩部具有发达的肌肉，因而高高隆起。棕熊体毛颜色多变，包括灰黑色、棕黑色、深棕色、棕红色、浅棕黄色及灰色，通常为棕褐色，四肢毛发颜色近黑色，偶见白化个体。

分布： 棕熊是全世界熊科动物中分布最广的物种，分布范围包括北半球亚欧大陆的大部分，和北美洲大陆北部与西部。在我国主要分布在黑龙江、吉林、辽宁、内蒙古、新疆、甘肃、青海、西藏、四川。

寿命： 野生条件下寿命约为30年

保护等级： 世界自然保护联盟（IUCN）红色名录无危物种，中国国家二级重点保护野生动物。

Chapter 3　第三章
>> Discovery Trail
发现熊猫

天之骄子，中华瑰宝，
生命奇迹，绿野遗踪。

The pride of the mountains, the national treasure of China, and the lost trace in green forests.

有人不太理解：既然我们大熊猫已经在地球上生存了至少800万年，为什么世界关注到我们的时间并不长呢？这是因为，我们主要栖息在海拔1500～3000米的高山竹林中，而且性情孤僻，喜欢独居，没有固定的居住地点，所以人们很难在野外见到我们，对我们缺乏足够的了解。在人类的史书上，我们在不同的地方、不同的故事里拥有各种各样的名字，我们身上总是笼罩着一层神秘色彩。因为好奇，人们还创想出了很多跟我们相关的传说故事。

1869年，一个法国人发现了我们，用科学方法对我们进行了命名，并把我们介绍给了全世界。我们奇特的生活习性和黑白相间的独特外表吸引了所有人的目光。人类为大自然的造物惊叹不已，"大熊猫热"由此席卷全球。

>> It might be confusing that even though we giant pandas have been living on earth for at least 8 million years, why didn't the world know about us for so long? This is because we mainly inhabit bamboo forests in the high mountains at an altitude of 1,500 to 3,000 meters, plus we are aloof and like to live alone without a fixed habitat. That helps explain why it is hard for humans to get a glimpse of us in the wild, and hence lack sufficient understanding about us. In history books, we have a variety of names in different places and in different stories, which has somewhat mystified us. To this end, many legends have been made up about us.

In 1869, a Frenchman discovered us, named us scientifically and introduced us to the world. Our strange habits and distinctive black and white appearance attracted the world's attention. We, the nature's creature, amazed you humans and gave rise to a world-sweeping "giant panda craze".

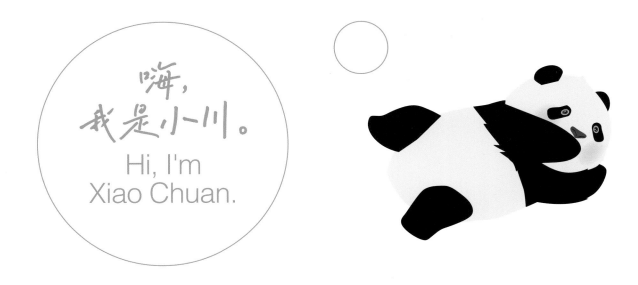

嗨，我是小川。
Hi, I'm Xiao Chuan.

>>

Chengdu
Giant Panda
Museum
Exhibition Hall 3
Discovery Trail

成都大熊猫博物馆 第三展厅 发现熊猫

Ancient Records

古籍里的记载

五花八门的名字
Various Names

Giant pandas have been living in China since ancient times with a wide distribution range. Many records of suspected panda species abound in ancient Chinese books.

In *The Book of Documents* and *The Classic of Poetry* compiled at the beginning of the Western Zhou Dynasty more than 3,000 years ago, the term "Pixiu" (a Chinese mythical hybrid creature) appeared, the first character of which was used to describe the mighty and valiant warriors. The *Shuowen Jiezi*, a Chinese dictionary compiled during the Eastern Han Dynasty recorded an animal called the Mo or tapir, with black and yellow fur, which was found in Shu (modern-day Sichuan Province). In the Northern Song Dynasty, the history book *Comprehensive Mirror in Aid of Governance* compiled by Sima Guang mentioned a benevolent beast called the "Zouyu". It was said that during wars, as long as one side raised the "Zouyu Flag", it would stand for the plea for peace.

Some believe that names such as Pixiu, Mo, Zouyu, white bear, black and white bear, mottled bear, bamboo bear, iron-eater mentioned in the ancient books all refer to pandas because pandas covered a much wider distribution range in ancient times than they do now, and people in different regions called them by different names.

However, as the Chinese nation is a multi-ethnic, multi-tribal melting pot formed after thousands of years of fusion, the ancestors of the various ethnic groups have their own totems, flags, legends, as well as dialects. In addition, the ancient texts are based on descriptions or notes from classics and commentaries instead of actual observation and scientific classification, which may lead to the confusion among names. Therefore, whether the animals recorded in these ancient books refer to giant pandas or not remains to be further verified.

大熊猫自古以来就生活在中国这片热土上，分布范围广泛。古籍里对疑似大熊猫这一物种的记载有很多。

3000多年前西周初年汇编的《尚书》和《诗经》里，就出现了"貔貅"这个名词，人们用"貔"比拟威武英勇的勇士；东汉时期编著的语文工具书《说文解字》里面记载了一种叫貘的动物，皮毛黑黄色，出没于蜀地；北宋司马光编撰的史书《资治通鉴》里提到了一种叫"驺虞"的仁兽，据说在打仗的时候，只要有一方举起"驺虞旗"，就代表求和。

有人认为，古书里的貔貅、貘、驺虞、白熊、黑白熊、花熊、竹熊、食铁兽等指的都是大熊猫，理由是大熊猫在古代的分布范围比现在大得多，不同地方的人对大熊猫有不同的叫法。

不过，因为中华民族是由多民族、多部落经过几千年混合、融合而成的，各族的先民都有自己的图腾、旗帜、传说以及方言，古籍中的文字记载都是基于经传的描述或注疏，而不是以实际观察和科学分类为根据，这就可能存在名称记载上的混淆。因此，这些古籍中记载的动物是否指的是大熊猫，还有待于进一步考证。

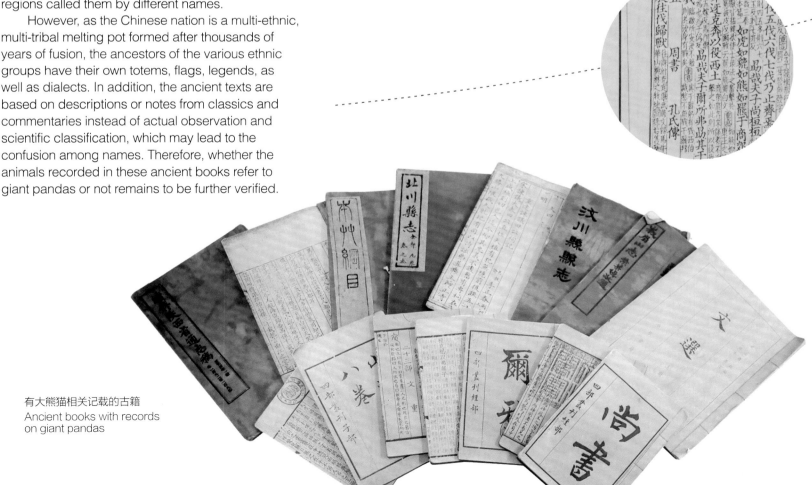

有大熊猫相关记载的古籍
Ancient books with records on giant pandas

人们以为我喜欢吃铁

>> Mistaken Cognition: I'm an Iron-eating Beast

秦汉时期，大熊猫族群仍怡然自得地生活在深山竹林中。它们有时会闯入人类居住区，摸进厨房舔舐铁锅中的盐分，甚至咬坏盛食物的铜铁器皿。它们这样做是为了补充饮食中缺乏的微量元素，但人类却误以为它们喜欢吃铁，所以把它们叫作"食铁兽"。

During the Qin and Han Dynasties, the giant panda community still lived comfortably in the bamboo forests. They sometimes intruded into human settlements, licked salt from iron pots in kitchens, and even chewed on copper and iron utensils that were used to hold food. The reason why they did so is to supplement the lack of trace elements, but humans mistakenly think they like to eat iron. Hence, they call the creature "iron-eater".

《康熙字典》收录了我
My name is included in the Kangxi Dictionary

In the *Kangxi Dictionary* compiled during the Kangxi period of the Qing Dynasty (1662~1722), pandas were called "Pixiu". Scholars have compared them with many real animals and concluded that only the giant panda is closest to the characteristics of the brave creature described in the *Kangxi Dictionary*.

在清朝康熙年间编撰的《康熙字典》里，大熊猫被称为"貔貅"。学者们对很多现实中的动物进行了比较论证，认为只有大熊猫最接近《康熙字典》里描述的貔貅的特征。

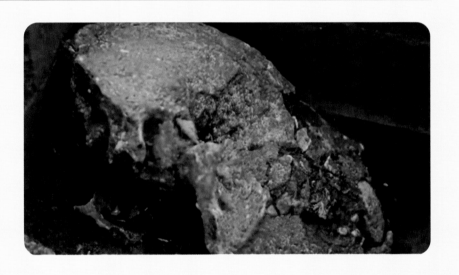

薄太后爱上大熊猫
Empress Dowager Bo's Adoration for Giant Pandas

The earliest archaeological evidence of human beings keeping giant pandas originated from a tomb in the Han Dynasty. It is believed that the owner of this tomb was Empress Dowager Bo of the Han Dynasty. Among the burial objects in this tomb are skulls of giant pandas, which means that people adored giant pandas more than 2,000 years ago.

人类最早饲养大熊猫的考古实证出自汉代的一个墓穴。据考证，这座墓的主人是汉代的薄太后。在这座墓穴的陪葬品中，人们发现了大熊猫的颅骨。看来，早在2000多年前，就有人爱上了大熊猫。

Science-based Discovery of the Giant Panda

科学发现大熊猫

历史上，在大熊猫野外栖息地附近生活的居民，根据大熊猫的体色、形貌、食性等特征，把它们称为"黑白熊""花熊""白熊""竹熊""食铁兽"等。大熊猫一直没有一个统一的名字，人们不知道这些顶着不同名头的"熊"和"兽"其实就是同一种动物。直到1869年，大熊猫这个物种才被来自法国的传教士阿尔芒·戴维发现，并被介绍到了全世界。

History has it that residents living near giant pandas' wild habitat called the species "black and white bears", "mottled bears", "white bears", "bamboo bears", "iron eaters", etc., according to their appearance like fur color, dietary preferences and other characteristics. There was not a unified name for the species, and people did not know those above-mentioned names actually referred to the same animal. It was not until 1869 that the species was named by the French missionary Armand David after discovery and introduced to the world at large.

阿尔芒·戴维
Armand David

Born in the mountains of western France, Armand David was a Lazarist missionary, but he displayed a strong interest in natural science as well. In 1850, he also systematically studied the natural sciences in Savona Seminary, Italy, and mastered the skills of making flora and fauna specimens. In 1861, a naturalist from the French National Museum of Natural History was planning a trip to China to complete his scientific research, and David was approved to go with him.

David first arrived in China in July 1862, and for the first few years he stayed in Beijing, conducting museum research in addition to his missionary work. He discovered and scientifically named the David's deer (*Elaphurus davidianus*), at the Nanhaizi Royal Hunting Gardens in the outskirts of Beijing.

In February 1866, David sailed from the French port of Marseilles for his second visit to China. In February 1869 he arrived at the Catholic Church of Dengchigou in Muping, Baoxing, Sichuan Province, where he served as the fourth priest and also collected flora and fauna specimens for the French National Museum of Natural History.

阿尔芒·戴维出生在法国西部山区，虽然是一位传教士，但是他一直对自然科学非常感兴趣。1850年，他还曾前往意大利的萨沃纳神学院系统学习了自然科学，掌握了制作动植物标本等方面的技能。1861年，法国国家自然历史博物馆的一位博物学家计划前往中国完成科学研究，戴维被批准一起前往。

戴维于1862年7月第一次抵达中国，在开始的几年里，他都待在北京，在传教之余进行博物学方面的研究。麋鹿就是他在北京郊区的皇家猎苑南海子中发现并进行科学命名的。

1866年2月，戴维从法国马赛港出发，第二次来到中国。1869年2月，他来到四川宝兴穆坪邓池沟天主教堂，在这里担任第四任神父，同时也为法国国家自然历史博物馆收集动植物标本。

阿尔芒·戴维
（Armand David）
1826~1900

阿尔芒·戴维日记中的大熊猫手绘图
Hand drawing of giant pandas in Armand David's diary

1869 年 3 月 11 日，戴维在野外考察时，受到当地一个姓李的猎人邀请，到他家里享用茶点。就在这里，他发现了挂在墙上的一张毛色黑白相间的兽皮，这引起了他极大的兴趣。

随后，他雇用了当地的猎户为他捕捉这种动物。3 月 23 日，猎户们捕捉到了一只活的幼年"黑白熊"，但它在运输途中死去了。戴维以高价买下了这只"黑白熊"。后来，他把这只"黑白熊"制作成了世界上第一具大熊猫标本。终于，在 4 月 1 日，猎户们为他带来了一只活的成年"黑白熊"。戴维根据大熊猫的形态特征，认为这是一种珍稀的熊类，于是将大熊猫的学名定为 *Ursus melanoleucus*，在拉丁文里的意思是黑白相间的熊。

从 1862 年到 1874 年，戴维三次深入中国内陆，在中国进行了 12 年的动植物资源考察，共发现了 189 个新物种，其中包括大熊猫、川金丝猴、羚牛、麋鹿、珙桐、大叶杜鹃、报春花等。他将这些新物种介绍给了全世界，在动植物标本收集、分类命名和生态描述等方面取得了令人瞩目的成绩。

On March 11, 1869, while on a field expedition, David was invited by a local hunter by the name of Li to enjoy refreshments at his home. It was there that he discovered a black and white animal pelt hanging on the wall, which piqued his interest.

He then hired local hunters to capture the animal for him. On March 23, the hunters captured a live juvenile "black and white bear," but it died in transit. David bought the bear for a high price, and he made it into the world's first giant panda specimen. Finally, on April 1, hunters brought him a live adult "black and white bear". Based on its morphological characteristics, David theorized that it is a rare species of bear, so he derived the scientific name of the giant panda as *Ursus melanoleucus* in Latin, which means the black-and-white bear.

From 1862 to 1874, David traveled three times into the remote interior of China, where he conducted 12-year flora and fauna expeditions, discovering a total of 189 new species, including the giant panda, golden snub-nosed monkey, the takin, David's deer, *Davidia involucrata*, *Rhododendron grande*, and *Primula malacoides*. He introduced these new species to the world and made remarkable achievements in the collection of flora and fauna specimens, taxonomic nomenclature, and ecological description.

19 世纪的法国马赛港，阿尔芒·戴维从这里出发前往中国
The port of Marseilles, France, in the 19th century, from which Armand David sailed for China

邓池沟天主教教堂
Dengchigou Catholic Church

阿尔芒·戴维在中国发现的部分物种
Some of the species Armand David found in China

大熊猫
Giant Pandas

大卫两栖甲
Amphizoa davidis

羚牛
Takin

珙桐
Davidia involucrata

麋鹿
David's deer

血雉
Blood pheasant

川金丝猴
Golden snub-nosed monkey

绿尾虹雉
Chinese monal

邓池沟天主教教堂内景
Inside Dengchigou Catholic Church

大熊猫有了学名
The Giant Pandas' Scientific Name

1870年，法国国家自然历史博物馆馆长阿方斯·米尔恩·爱德华兹对戴维寄回法国的"黑白熊"标本重新进行了研究。他认为"黑白熊"更接近在中国西藏发现的小熊猫，"黑白熊"和小熊猫有很多相似之处，只是体形差别有点儿大。所以，爱德华兹把戴维给"黑白熊"归类的属名 Ursus（熊属）改为 Ailuropoda（猫熊属），种名 melanoleuca 保留，确定了大熊猫这一物种的学名 Ailuropoda melanoleuca，在拉丁文里的意思是黑白相间的猫熊。

对于大熊猫的学名，分类学家们曾有很多争论。后来，经过反复研究讨论，他们最终还是认为 Ailuropoda melanoleuca（黑白相间的猫熊）最合适，这一学名也就一直被使用并延续了下来。

In 1870, Alphonse Milne Edwards, director of the French National Museum of Natural History, re-examined the "black and white bear" specimen that David had sent back to France. He discovered that the species was closer to the red panda found in Xizang, China and that they share some similarities except for size. Therefore, Edwards changed the genus name *Ursus* to *Ailuropoda* but retained the species name *melanoleuca* to determine the scientific name of the species as *Ailuropoda melanoleuca* in Latin, which means black-and-white cat bear.

The scientific name of giant pandas has been controversial among taxonomists. After repeated studies and discussions, they finally decided that *Ailuropoda melanoleuca* (black-and-white cat bear) was the most appropriate, and this scientific name is still used today.

法国国家自然历史博物馆历史悠久，馆藏丰富，是一座举世闻名的自然历史博物馆。阿尔芒·戴维寄回法国的大熊猫标本就保存在这里。

The French National Museum of Natural History, a world-renowned natural history museum, boasts a long history and abundant collections. It houses the giant panda specimen sent back to France by Armand David.

从"猫熊"到"熊猫"
>> From "Cat Bear" to "Panda"

The origin of the Chinese name of the giant panda has different stories. One popular version goes: in 1869, David discovered the giant panda and named it *Ursus melanoleucus* in Latin (cat bear in English), which translates to "maoxiong" (cat bear) in Chinese.

Giant pandas were first exhibited within China in 1939 at the Beibei People's Park (now Beibei Park) in Chongqing. The name giant panda was written in both English and Chinese on the name tag at that time. To remain consistent with the English writing style, the Chinese name was also written in a left-to-right format. However, since the Chinese language at that time was usually written from right to left, tourists read it in reverse, so they read "Xiongmao" instead of "Maoxiong". To distinguish between the two types of pandas, people called the panda found in Tibet (Xizang), China in 1825 the red panda, and those found by Armand David in 1869 the giant panda. Since then, the name giant panda or "Daxiongmao" in Chinese are still in use today.

关于大熊猫中文名字的由来,有着不同的说法。比较流行的一种说法是:1869 年,戴维发现大熊猫后,将其拉丁学名定为 *Ursus melanoleucus*,英语为"Cat Bear",翻译为中文就是"猫熊"。

1939 年,大熊猫第一次在中国国内展出,地点是重庆北碚平民公园(今北碚公园)。当时展出的名字标牌上分别用中、英文书写了大熊猫的名字。为了和英语的书写方式保持一致,中文名也采用了从左往右的书写形式。由于当时中文的读写习惯是从右到左,因此游人们都将"猫熊"读成了"熊猫"。后来,为了区别两种熊猫,人们就把 1825 年在中国西藏发现的熊猫称为小熊猫或红熊猫,而把 1869 年阿尔芒·戴维发现的熊猫称为大熊猫。此后,"大熊猫"这一名称便约定俗成地流传了下来。

Giant Panda Going Global

大熊猫走向世界

Su Lin: The First Live Giant Panda to Travel Out of China

苏琳：走出中国的第一只活体大熊猫

After David sent the giant panda specimen he found in Baoxing County, Sichuan Province, back to Paris, France, for exhibition, this mysterious species grabbed widespread attention in the Western zoological community. Since then, the species had undergone hard times - Many Westerners, for the causes of expedition, exploration, hunting, and missionary, traveled thousands of miles to the mountainous areas of western China, collected and hunted giant pandas for making specimens to be displayed in Western museums. Even worse, live pandas were taken out of China, and American Ruth Harkness was the "pioneer".

Ruth Harkness was a clothing designer, and her husband, William Harkness, was an avid explorer. A few weeks after their marriage, William traveled with his classmates from the United States to China looking to capture live giant pandas as he aspired to be the first person to ship a live giant panda out of China. However, shortly after arriving in Shanghai, his classmates returned U.S. William spent a year and a half in China without even catching a glimpse of a giant panda and eventually died of cancer in Shanghai.

In grief, Ruth was resolved to fulfill her husband's wishes and traveled to China. Upon arrival, she quickly organized an expedition team. On November 9, 1936, they found a panda cub, about a few months old, in a dry hole in a tree in Caopo Township, Wenchuan, which Ruth named Su Lin. With the help of a friend, Ruth then bribed her way through Chinese customs and obtained an export license disguised as "one dog, 20 US dollars". That was how Ruth brought the cub back to the United States.

戴维将在四川省宝兴县发现的大熊猫的标本寄回法国巴黎展出后，这一神秘的物种引起了西方动物学界的广泛关注。此后，大熊猫遭遇了一段艰难岁月：很多西方人先后以考察、探险、狩猎、传教等名义，不远万里来到中国，深入中国西部山区，收集、猎杀大熊猫，并制作标本供西方博物馆展出。更有甚者，将活体大熊猫带出了中国。美国人露丝·哈克尼斯就是其中第一人。

露丝·哈克尼斯是一个服装设计师，她的丈夫威廉·哈克尼斯是一个狂热的探险家。威廉与露丝结婚几周后，便与他的同学从美国出发来到中国，计划猎捕活体大熊猫。他立志要成为将活着的大熊猫运出中国的第一人。但是，抵达上海后不久，他的同学就回国了。威廉在中国待了一年半，却连大熊猫的影子都没有见到，最后因癌症在上海病逝。

露丝在悲痛之余决心完成丈夫的遗愿，所以她也来到了中国。她在抵达中国后迅速组织了一支探险队。1936年11月9日，他们在汶川草坡乡一个干枯的树洞中发现了一只出生才几个月的熊猫幼仔。露丝给它取名"苏琳"。随后，露丝在朋友的帮助下买通中国海关，得到了"狗一只，20美元"的出口许可。就这样"苏琳"被露丝带回了美国。

露丝·哈克尼斯和"苏琳"
Ruth Harkness and Su Lin

1981年上映的中日合拍动画电影《熊猫的故事》就取材于苏琳的经历。

The Chinese-Japanese animated film *The Story of Panda-Taotao*, released in 1981, was based on Su Lin's experiences.

熊猫世界

After coming to the United States, Su Lin was widely reported by the press and was all the rage. In 1937, it was sold by Ruth to the Brookfield Zoo in Chicago for $8,750. At the zoo, Su Lin immediately became a super star and set off a "giant panda craze" across the United States.

However, the high did not last long, Su Lin was not adapted to the climate and food in the U.S., and constantly fell ill. Unfortunately, on April 1, 1938, Su Lin died of pneumonia at just over a year old.

It was the first giant panda in modern scientific history to be in the public eye, and the first live giant panda which was taken out of China and reached the U.S. Its untimely death was great tragedy.

来到美国后的"苏琳"被媒体大肆报道，受到人们的疯狂追捧。1937年，露丝以8750美元的价格，将"苏琳"卖给了芝加哥布鲁克菲尔德动物园。"苏琳"到了动物园后立即成为超级动物明星，并在美国掀起了一股"大熊猫热"。

不过，好景不长，"苏琳"由于不适应美国的气候和食物，不断生病。1938年4月1日，1岁多的苏琳因感染肺炎死亡。

"苏琳"是现代科学史上第一只走进公众视野的大熊猫，也是第一只被带出中国、登陆美国的活体大熊猫，它的英年早逝让人们深感惋惜。

"姬姬"：世界自然基金会形象大使
Ji Ji: WWF Ambassador

1955年，来自宝兴县盐井乡和平村的大熊猫"平平"和来自硗碛乡的大熊猫"碛碛"落户北京动物园，成为第一批入住北京动物园的大熊猫。

1957年正值苏联十月革命胜利40周年，苏联政府代表团出访中国时在北京动物园看到了憨态可掬的大熊猫，向中国政府提出了希望获赠大熊猫的请求。于是，"雌性"大熊猫"平平"和"雄性"大熊猫"碛碛"被作为国礼赠送给了苏联政府。

两只大熊猫在莫斯科动物园里生活了半年之久，"平平"却没有任何怀孕的迹象。莫斯科动物园再次给它们做了性别鉴定，结论是：它们俩应该都是雄性。于是"碛碛"被送回国，另一只同样来自宝兴的大熊猫"安安"作为"平平"的配偶被送到了苏联。有意思的是，被送回的"碛碛"才是真正的雌兽，而"安安"可是正儿八经的雄兽。就这样，定下"娃娃亲"的"平平"和"碛碛"被拆散了。

"碛碛"被送回北京动物园后，改名为"姬姬"。1958年，奥地利动物商人海尼·德默尔来到中国，用一批来自非洲的动物（三只长颈鹿、两只犀牛、两只河马、两只斑马）与北京动物园交换了一只大熊猫，这只大熊猫就是"姬姬"。在此之前，德默尔与美国一家动物园签署了协议，约定将以2.5万美金

In 1955, the giant panda Ping Ping from Heping Village, Yanjing Township, Baoxing County and the giant panda Qi Qi from Qiaoqi Township settled in Beijing Zoo, becoming the first giant pandas to stay in Beijing Zoo.

1957 marked the 40th anniversary of the victory of the October Revolution of the Soviet Union, and a delegation of the Soviet government visited China and saw the pandas at the Beijing Zoo. They put forward a request to the Chinese government to be gifted with giant pandas. In response, the female giant panda Ping Ping and the male one Qi Qi were given to the Soviet government as a present.

The two giant pandas had been living in the Moscow Zoo for half a year, but Ping Ping did not show any signs of pregnancy. The Moscow Zoo conducted sex identification again and concluded that the two were both male. So Qi Qi was sent back to China and another giant panda An An from Baoxing was chosen to be sent to the Soviet Union. Interestingly, after Qi Qi returned, she was found to be female, while An An was male. Given that, Ping Ping and Qi Qi, which were originally set for "betrothal", were separated.

After Qi Qi's trip back to the Beijing Zoo, she was renamed Ji Ji. In 1958, an Austrian animal trader Heini Demmer came to China and exchanged a batch of animals from Africa (3 giraffes, 2 rhinoceroses, 2 hippopotamuses, and 2 zebras) with the Beijing Zoo for a giant panda, which turned out to be Ji Ji. Prior to this, Demmer signed an agreement with a zoo in the United States that a giant panda would be sold to it for 25,000 dollars. However, for various reasons, the entry of giant pandas was prohibited in the United States. Demmer ended up having no choice but to take Ji Ji to Europe on a travelling exhibition.

At the end of 1958, Ji Ji came to the London Zoo in England and ultimately settled down, ending her displaced life. As the first giant panda to arrive in Europe after World War II, Ji Ji was warmly welcomed by the locals.

In 1961, the World Wildlife Fund (later renamed the World Wide Fund for Nature, WWF for short) was founded. This organization dedicated to the protection of the global biodiversity and living environment realized that an influential logo can be understood by all without language barriers, and thus decided to use Sir Scott's design of the giant panda for the flag and emblem of the WWF. The giant panda image became the symbol of the organization and a well-known icon of the global conservation movement, which is actually modeled on the giant panda Ji Ji.

In a sense, giant pandas began to take the world by storm thanks to Ji Ji. Her charm has driven the cause of wildlife conservation around the world. The image and clout of the giant panda family represented by Ji Ji has thus taken root throughout the globe.

的价格向其出售一只大熊猫。但是，由于种种原因，美国禁止大熊猫入境。无奈之下，德默尔只好带着"姬姬"在欧洲巡展。

1958年末，姬姬来到了英国伦敦动物园定居下来，从此结束了颠沛流离的生活。作为二战后第一只到达欧洲大陆的大熊猫，"姬姬"受到了当地民众的热烈欢迎。

1961年，世界野生动植物基金会（后更名为"世界自然基金会"，英文缩写为WWF）宣告成立。这个致力于保护世界生物多样性及生物生存环境的机构意识到，具有影响力的组织标志可以克服所有语言上的障碍，于是决定将斯科特爵士设计的大熊猫图案用于WWF的会旗和会徽。大熊猫的形象就此成为该组织的象征，成为全球自然保护运动的著名标志。这一标志的原型就是大熊猫"姬姬"。

从某种意义上说，大熊猫开始风靡世界，就是从"姬姬"开始的。"姬姬"的魅力，深深影响了世界野生动物保护的走向和进程。"姬姬"所代表的大熊猫家族的形象与影响力，就这样在全世界生根发芽。

世界自然基金会标志的变迁
Changes to the WWF logo

Giant Pandas' Conservation and Research

大熊猫保护研究

熊猫是中国的国宝，也是世界级的珍稀物种。中国政府非常重视大熊猫的保护研究工作，从 20 世纪 50 年代起，一大批中、外学者先后参与到大熊猫的科学研究与保护工作中来，在大熊猫种群调查、繁殖育幼、遗传管理、疾病防治、栖息地保护等方面做出了巨大的贡献。正是有了他们的努力和付出，人们才逐渐揭开了大熊猫野外生活的神秘面纱，对大熊猫的种群数量和栖息地范围也有了越来越精确的记录。同时，圈养大熊猫种群质量的提升，也为日后野外种群的恢复与繁衍奠定了坚实的基础。

As a national treasure of China, the giant panda is also a rare species in the world. The Chinese government attaches great importance to the protection and research of giant pandas. Since the 1950s, large throngs of Chinese and foreign scholars have been engaged in the scientific research and protection of the species and have made tremendous contributions to the investigation of giant panda populations, breeding and raising of cubs, genetic management, disease control, habitat protection, and other aspects. It is through their efforts and dedication that the mystery of the giant panda's life in the wild has been gradually unveiled, and increasingly accurate records of their population size and habitat range have been maintained. Additionally, the improved quality of the captive giant panda population has also laid a solid foundation for the recovery and reproduction of the wild population in the future.

The Pioneer of Giant Pandas' Conservation and Research

In 1973, Premier Zhou Enlai convened a symposium of researchers from Sichuan, Gansu, and Shaanxi provinces where he asked for a survey of rare animals, mainly giant pandas. In 1974, Hu Jinchu, at the age of 45, was commissioned to station in Wolong, Sichuan. He established the Sichuan Rare Animal Resource Survey Team and led China's first field survey of giant pandas. As the leader of this survey, Hu visited all the giant panda habitats in Sichuan.

In 1980, China established the China Conservation and Research Center for Giant Panda in Wolong in cooperation with the WWF. Outstanding wildlife researcher and famous zoologist George Schaller and several other foreign experts, together with Chinese scientist representative Hu Jinchu and his research team, jointly established the Wuyi Hut, the world's first field ecological observatory for giant pandas. Since then, Hu Jinchu and Schaller have led Chinese experts in engaging in wild giant panda research, collecting and compiling a large quantity of data on wild giant pandas.

In the days of tracking and monitoring wild giant pandas, they trapped wild giant pandas, anesthetized them, and installed radio collars before releasing them into the wild for tracking and monitoring. This allowed them to obtain a wealth of first-hand information on the behaviors and habits of wild giant pandas, such as feeding, estrus, courtship, reproduction, and pathology, which paved the way for conservation in the years to come.

1973年，周恩来总理召集四川、甘肃、陕西三省召开座谈会，要求开展以大熊猫为主的珍稀动物调查。1974年，45岁的胡锦矗受命进入四川卧龙，组建了四川省珍稀动物资源调查队，组织和领导了全国第一次大熊猫野外调查研究。作为此次调查的队长，胡锦矗考察了四川所有的大熊猫栖息地。

1980年，中国与世界野生动植物基金会（后更名为"世界自然基金会"，英文简称WWF）合作，在卧龙建立了"中国保护大熊猫研究中心"。杰出的野生动物研究专家、著名动物学家乔治·夏勒以及其他几位外国专家，与中方科学家代表胡锦矗及其研究团队，共同建立了世界上第一个大熊猫野外生态观测站——"五一棚"。此后，胡锦矗与夏勒带领中外专家们一起参与到野生大熊猫研究工作中，收集整理了大量关于野生大熊猫的数据资料。

在追踪和监测野生大熊猫的日子里，他们通过诱捕野生大熊猫，将其麻醉后戴上无线电颈圈再放归野外进行跟踪监测的方式，获取了大量野生大熊猫取食、发情、求偶、生育、病理等行为和生活习性的第一手资料，为此后开展的大熊猫保护工作奠定了基础。

中国大熊猫保护研究中心
China Conservation and Research Center for the Giant Panda

胡锦矗
中国大熊猫研究泰斗

胡锦矗是西华师范大学教授、著名动物学家，中国保护大熊猫研究中心第一任主任，中国大熊猫研究的标志性和泰斗级人物，国际公认的大熊猫生态生物学研究奠基人。

他于1974年组织和领导了全国第一次大熊猫野外调查研究；1978年在卧龙建立了世界上首个野生大熊猫生态观测站——"五一棚"野外观测站；1980年担任中方专家组组长，与世界自然基金会（WWF）开展国际合作，运用无线电遥测技术跟踪野生大熊猫。

他荣获了多项国家级科学技术奖和"全国优秀科技工作者"等荣誉；2007年获世界自然基金会（WWF）自然保护贡献奖，2008年获"斯巴鲁野生动物保护奖"，2019年获"大熊猫科学研究和保护终身成就奖"。

Hu Jinchu
A leading researcher of giant pandas in China

Hu Jinchu is a professor at China West Normal University, a famous zoologist, the first director of the China Conservation and Research Center for Giant Panda. As an iconic mind in China's giant panda research, he is also internationally recognized as the pioneer of research into the ecology and biology of the giant panda.

He organized and led China's first field survey of giant pandas in 1974 and established the world's first ecological observatory of wild giant pandas in Wolong in 1978, the Wuyi Hut Field Observatory. In 1980, as the head of the Chinese expert group, he handled international cooperation with the WWF in tracking wild giant pandas via radio telemetry.

He has won many scientific and technological awards and honors such as the National Outstanding Scientific and Technological Worker. In 2007, he was awarded the WWF Award for Contribution to Nature Conservation, in 2008, the Subaru Wildlife Conservation Award, and in 2019, the Lifetime Achievement Award for Scientific Research and Conservation of Giant Pandas.

George Schaller
A dedicated researcher of giant panda behaviors

George Schaller is a prominent American field conservation biologist, naturalist, conservationist, and author. Dedicated to wildlife conservation and research, he has conducted extensive zoological research in Africa, Asia, and South America, and has been recognized by *Time* as one of the world's 3 most outstanding scholars of wildlife research. He is also the first Western scientist to be commissioned to work for the WWF in China.

In 1980, Schaller began his behavioral studies of giant pandas in the Wolong Nature Reserve in Sichuan Province. For his contributions to conservation, Schaller was awarded the WWF's Gold Medal, the International Cosmos Award (Japan), and the Taylor Prize for Environmental Achievement (USA). In 2019, he was awarded the Lifetime Achievement Award for Scientific Research and Conservation of Giant Pandas at the China (Sichuan) Giant Panda International Eco-Tourism Festival.

乔治·夏勒
致力于大熊猫行为学研究

乔治·夏勒是美国著名的野外保护生物学家、博物学家、自然保护主义者和作家。他致力于野生动物的保护和研究，在非洲、亚洲、南美洲广泛开展动物学研究，曾被美国《时代周刊》评为世界上三位最杰出的野生动物研究学者之一，也是第一个受委托在中国为世界自然基金会（WWF）开展工作的西方科学家。

1980年，夏勒开始在四川卧龙自然保护区对大熊猫进行行为学研究。由于对自然保护的贡献，夏勒荣获了世界自然基金会颁发的"金质勋章"和"国际宇宙奖（日本）""泰勒环境成就奖（美国）"等荣誉。2019年在中国（四川）大熊猫国际生态旅游节上，他被授予"大熊猫科学研究和保护终身成就奖"。

五一棚：世界上第一个大熊猫野外观测站
Wuyi Camp: The World's First Giant Panda Field Observatory

1978年，胡锦矗等新中国第一代大熊猫研究专家在四川卧龙自然保护区内大熊猫经常出没的一座山上搭起帐篷，建立了世界上第一个大熊猫野外生态观测站，在该区域内首次开展大熊猫等动植物生态科研工作。因为搭建帐篷的地方离水源地有五十一步台阶，故取名"五一棚"。

"五一棚"海拔2520米，由山下到"五一棚"要沿着"之"字形山路攀爬约两小时。"五一棚"初建时仅由帆布帐篷组成，棚内使用煤油灯照明，烧枯柴做饭，再加上几张木头钉的简易床，没有任何文化娱乐设施，一切生活用品都靠人力背上山，条件极为艰苦。

1986年后，"五一棚"才通了电，生活和科研条件有了很大改善。在此期间，研究者们获得了大量有关大熊猫的第一手资料，这也是迄今为止人们获取的最具权威性的野生大熊猫资料。

"五一棚"所开创的事业，为后来轰轰烈烈、旷日持久的大熊猫保护运动开了先河。随着时代的发展，"五一棚"经过20世纪70年代的兴起、80年代的辉煌、90年代的持续和21世纪的更新，当年的小小帐篷已经发展成一个有着完备设施的工作站。"五一棚"见证了一代又一代大熊猫科研保护人员的努力和他们取得的成果，它是中国大熊猫保护研究事业的一块重要基石。

In 1978, in a mountain where giant pandas are often seen in the Sichuan Wolong Nature Reserve, Hu Jinchu and other pioneering giant panda research experts established the world's first field ecological observatory for giant pandas where they conducted the first ecological scientific research on giant pandas and other plants and animals in the area. Because the hut were erected 51 (wuyi in Chinese) steps away from the water source, it was named the Wuyi Camp.

The place is 2520 meters above sea level. It takes about 2 hours to travel in a zigzag route from the bottom of the mountain to the Wuyi Camp. The conditions were extremely harsh when it was first built. Its initial layout consisted only of canvas tents, kerosene lamps for lighting, wood for cooking, and a few simple beds nailed by the wood, without any other facilities. All daily necessities had to be carried up the mountain.

It was only after 1986 that electricity was provided, greatly improving living and research conditions. During this period, researchers obtained a large amount of first-hand information about giant pandas, which is still proved to be the most authoritative information about wild giant pandas that people has been collected so far.

20世纪80年代
In the 1980s

20世纪90年代
In the 1990s

What has been achieved in Wuyi Camp has paved the way for the later and longer-lasting campaign of giant panda conservation. With the development of the times, the place has been transformed into a workstation with complete facilities through the rise of the 1970s, the splendor of the 1980s, the continuity of the 1990s, and the renewal in the 21st century. It has witnessed the efforts and achievements of generations of giant panda researchers and protectors, acting as an important cornerstone of China's giant panda conservation and research endeavors.

在简陋的帐篷外工作
Working outside the humble tent

"五一棚"最初的模样
The initial Wuyi Camp

2000 年
In 2000

现在
Now

Messenger of Friendship

友谊的使者

人见人爱、花见花开的大熊猫走出中国西部山林，登上世界大舞台后，征服了广大海外民众，成为动物界的"顶流"。在很多人心目中，"大熊猫"几乎成了中国的代名词。从喜爱大熊猫开始，不同种族不同国度的人们慢慢对中国和中国文化产生了兴趣。大熊猫用自己独特的魅力，成功搭建起了一座中国与世界友好交往的桥梁。

The widely beloved giant pandas have emerged out of the mountains in western China to the world stage, conquering the rest of the world and wining the largest fan base in the animal world. To many, giant pandas have almost become synonymous with China. Thanks to their fondness of giant pandas, people of different races and countries have gradually become interested in China and Chinese culture as well. It can be said that giant pandas have successfully built a bridge of friendly communication between China and the world at large with their unique charm.

动物大使
>> The Animal Ambassador

The precious and rare species is a symbol of peace and friendship. Since ancient times, they have been sent abroad many times as part of the effort to facilitate foreign exchange. According to Japan's *Royal Annals*, Wu Zetian, the only female emperor in Chinese history, gave two "white bears" (giant pandas) to the Emperor of Japan as a gift, which is the earliest record of giant pandas serving as a "friendship ambassador" in international exchanges.

From 1957 to 1983, China presented 24 giant pandas to 9 countries: the USSR, the DPRK, the United States, Japan, France, the United Kingdom, West Germany, Mexico and Spain. These "messengers of exchanges" were warmly welcomed by the people of the recipient countries and greatly contributed to developing China's friendly ties with foreign countries.

Out of the need to protect endangered animals, China ceased to voluntarily give giant pandas to foreign countries after 1983, and instead by sending giant pandas on short-term loans to foreign countries for exhibition tours. Giant pandas were warmly welcomed by the government and people of each country they visited, which promoted the friendship between the Chinese people and those of other countries. However, the loan program was abolished in 1993.

Since the 1990s, China has conducted international scientific research cooperation on giant pandas with a number of countries. Giant pandas have gone abroad as "ambassadors of scientific research cooperation", which has not only helped researchers overcome a number of difficulties in the field of giant panda conservation and breeding, but also acted as a catalyst for Chinese-foreign cultural exchanges and the development of global ecological conservation education.

大熊猫珍贵且稀少，是和平友好的象征，从古至今，它们曾多次被送到国外，肩负着对外交流的使命。根据日本《皇家年鉴》的记载，中国历史上唯一的女皇帝武则天曾将两只"白熊（大熊猫）"作为国礼赠送给日本天皇。这是大熊猫在国际交往中担任"友谊使者"的最早记录。

1957年~1983年，中国先后向苏联、朝鲜、美国、日本、法国、英国、西德、墨西哥和西班牙9个国家赠送了24只大熊猫，这些"交流使者"受到了各国人民的极大欢迎，对发展我国的对外友好关系做出了巨大贡献。

出于保护濒危动物的需要，1983年后中国取消了向国外无偿赠送大熊猫的做法，改用短期借展的方式将大熊猫送往国外巡展。大熊猫每到一个国家，都受到该国政府和人民的热烈欢迎，增强了中国人民与世界人民之间的友谊。不过，借展的方式在1993年也被取消了。

20世纪90年代以来，中国先后与多个国家开展了大熊猫国际科研合作，大熊猫以"科研合作大使"的身份去到国外，不仅帮助研究者们攻克了多项大熊猫保护繁育领域的难关，还对中外文化交流和全球生态保护教育的发展起到了促进作用。

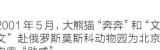

1972年11月，大熊猫"康康"和"兰兰"在日本东京上野动物园和观众见面。
In November 1972, giant pandas Kang Kang and Lan Lan met visitors at Ueno Zoo in Tokyo, Japan.

2019年4月，大熊猫"星二"和"毛笋"在丹麦哥本哈根动物园与游客见面。
In April 2019, giant pandas Xing Er and Mao Sun met visitors at Copenhagen Zoo in Denmark.

2001年5月，大熊猫"奔奔"和"文文"赴俄罗斯莫斯科动物园为北京申奥"助威"。
In May 2001, giant pandas Ben Ben and Wen Wen traveled to the Moscow Zoo in Russia to support the Beijing's Olympic bid.

大熊猫的远行足迹
Giant Panda Global Footprints

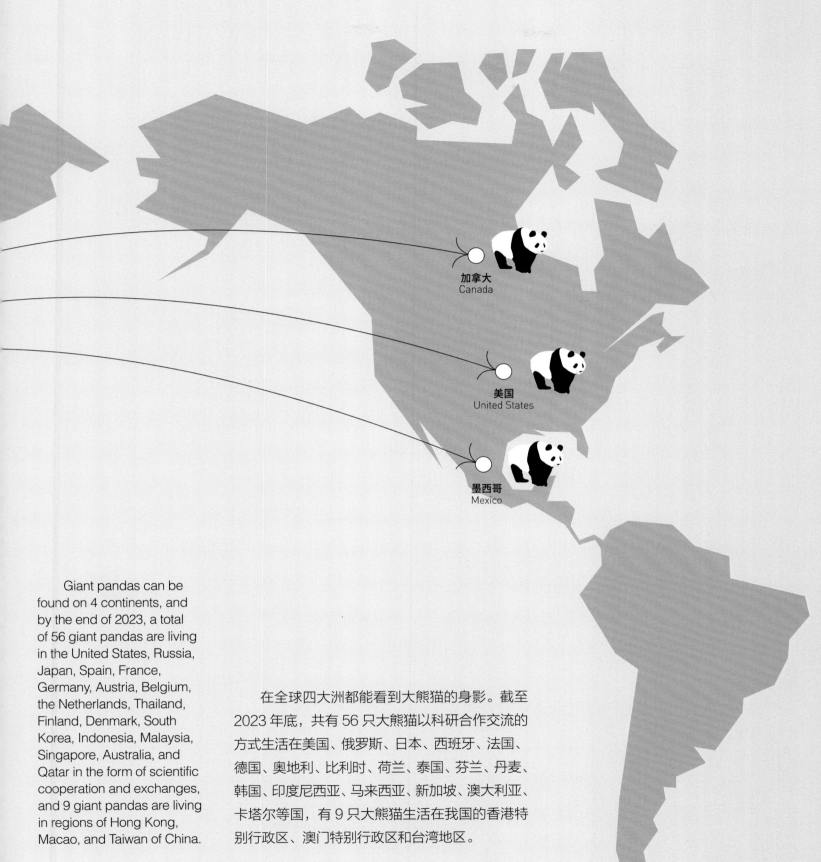

Giant pandas can be found on 4 continents, and by the end of 2023, a total of 56 giant pandas are living in the United States, Russia, Japan, Spain, France, Germany, Austria, Belgium, the Netherlands, Thailand, Finland, Denmark, South Korea, Indonesia, Malaysia, Singapore, Australia, and Qatar in the form of scientific cooperation and exchanges, and 9 giant pandas are living in regions of Hong Kong, Macao, and Taiwan of China.

在全球四大洲都能看到大熊猫的身影。截至2023年底，共有56只大熊猫以科研合作交流的方式生活在美国、俄罗斯、日本、西班牙、法国、德国、奥地利、比利时、荷兰、泰国、芬兰、丹麦、韩国、印度尼西亚、马来西亚、新加坡、澳大利亚、卡塔尔等国，有9只大熊猫生活在我国的香港特别行政区、澳门特别行政区和台湾地区。

各具特色的海外大熊猫馆
Overseas Giant Panda Houses with Their Own Character

你气势壮观,我富丽堂皇!你用四川味亲近它?那我就拿八卦图作外观!爱它,就要千方百计对它好。为了迎接来自中国的大熊猫,国外动物园都会特别建造一座大熊猫馆。这些大熊猫馆都极具特色,充分显示了全世界人民对大熊猫的喜爱。让我们来看看这几座最特别的海外大熊猫馆吧!

Magnificent and opulent houses and those built with Sichuan style or Bagua-trigram layouts are all designed to cater to giant panda needs as a way to show how deeply they adore the species. To welcome giant pandas, foreign zoos from all over the world build special panda houses with unique features to show their affection. Let's take a look at some of these most distinctive overseas giant panda houses.

德国柏林动物园大熊猫馆:超级豪华别墅
Giant Panda House at Berlin Zoo, Germany:
Super Luxury Villa

德国柏林动物园是德国历史最悠久的动物园,自 1844 年开放,已有 173 年的历史。它位于柏林市中心,占地 35 公顷,是德国最大的动物园,也是世界上展出动物种类最多的动物园之一。柏林动物园大熊猫馆占地约 5480 平方米,馆内种植了大量竹类植物。馆内温控系统等硬件配套设施完善,堪称大熊猫馆里的超级豪华别墅!

2017 年,时值中德建交 45 周年,成都大熊猫繁育研究基地的大熊猫"梦梦"与"娇庆"以科研合作交流的身份来到柏林动物园。2019 年,"梦梦"在这里生下了双胞胎"梦圆"和"梦想"。

Opened to the public in 1844, the Berlin Zoo is the oldest zoo in Germany with a history of 173 years. Located in the center of Berlin, it covers an area of 35 hectares and is the largest zoo in Germany as well as one of the zoos with the largest number of animal species on exhibition in the world. The giant panda house covers an area of about 5,480 square meters with a large number of bamboo plants. It is a super luxury villa for giant pandas equipped with the perfect facilities such as a temperature control system.

In 2017, on the occasion of the 45th anniversary of the establishment of diplomatic relations between China and Germany, giant pandas Meng Meng and Jiao Qing from the Chengdu Research Base of Giant Panda Breeding came to the Berlin Zoo in the capacity of scientific research cooperation and exchange. In 2019, Meng Meng gave birth to twins named Meng Yuan and Meng Xiang.

柏林动物园大熊猫馆
Giant Panda House at Berlin Zoo

大熊猫"星二""毛二"出访丹麦,入住哥本哈根动物园

Giant pandas Xing Er and Mao Er visited Denmark and settled in Copenhagen Zoo

Giant Panda House at Copenhagen Zoo, Denmark:
Design Inspired by Tai Chi Patterns

Founded in 1859, Copenhagen Zoo in Denmark is one of the oldest zoos in Europe and the most visited in Denmark. In 2014, during Denmark Queen Margrethe Ⅱ's stay in China, she put forward her wish to carry out scientific research and cooperation on giant pandas with China. In 2017, giant pandas Xing Er and Mao Er from the Chengdu Research Base of Giant Panda Breeding visited Denmark and settled in Copenhagen Zoo.

The shape of the Denmark Copenhagen Zoo Giant Panda House is similar to the yin and yang of Bagua trigrams in tai chi patterns. Its design is inspired by the Chinese Tai Chi symbols and aims to provide giant pandas with a living environment that is closest to nature.

丹麦哥本哈根动物园大熊猫馆:设计灵感来自太极图案

丹麦哥本哈根动物园创立于1859年,是欧洲最古老的动物园之一,也是丹麦接待游客最多的动物园。2014年,丹麦女王玛格丽特二世访问中国期间,向中国提出开展大熊猫科研合作的愿望。2017年,来自成都大熊猫繁育研究基地的大熊猫"星二""毛二"出访丹麦,入住哥本哈根动物园。

丹麦哥本哈根动物园大熊猫馆的外形类似太极图案里的阴阳八卦。它的设计灵感来自中国的太极符号,目的在于给大熊猫提供最接近自然的生活环境。

法国博瓦勒野生动物园大熊猫馆：中国园林专家参与修建

法国博瓦勒野生动物园占地面积 30 公顷，是欧洲最大的动物园之一。

博瓦勒野生动物园为了给旅居大熊猫营造舒适的居所，仿照中国古典园林的样式为它们建设新家，还特意聘请了 28 名中国古典园林专家前往当地参与项目修建。博瓦勒野生动物园的大熊猫馆对大熊猫故乡四川的自然生态环境进行了模拟，为旅居大熊猫营造出了回家的感觉。2012 年，成都大熊猫繁育研究基地的大熊猫"圆仔"和"欢欢"来到博瓦勒野生动物园。2017 年，"欢欢"在这里生下了儿子"圆梦"。2021 年 8 月 2 日，"欢欢"又产下一对双胞胎，取名"欢黎黎""圆嘟嘟"。

Giant Panda House at ZooParc de Beauval, France: Chinese Garden Experts Contributed to Its Construction

ZooParc de Beauval in France is one of the largest zoos in Europe, covering an area of 30 hectares.

To create a comfortable environment for the sojourning giant pandas, the park built a new home for them in the style of Chinese classical gardens, and specially hired 28 Chinese classical garden experts to participate in the project. The panda house in the park simulates the natural ecological environment of Sichuan, the hometown of the giant pandas, to make them feel at home. In 2012, giant pandas Yuan Zai and Huan Huan from the Chengdu Research Base of Giant Panda Breeding traveled to the park. In 2017, Huan Huan gave birth to a son, Yuan Meng. On August 2, 2021, Huan Huan gave birth to twins named Huan Lili and Yuan Dudu.

博瓦勒野生动物园大熊猫馆
Giant Panda House at ZooParc de Beauval

Giant Panda House at the Ueno Zoo in Tokyo, Japan:
A Natural Environment Modeled After Sichuan Province

Opened to the public in 1882, the Ueno Zoo in Tokyo, Japan, covers an area of about 14 hectares with about 420 species of animals on exhibition, among which giant pandas are the most popular.

In 2011, giant pandas Bi Li and Xian Nv from the China Conservation and Research Center for Giant Panda moved into the Ueno Zoo's Giant Panda House. After collecting names from the public, the park named the two as Li Li and Zhen Zhen.

The Ueno Zoo's Panda Forest of the Giant Panda House is modeled after the natural landscape of Sichuan, the hometown of the giant pandas, such as trees, rocks, and water flow. Some of the spaces inside are open-plan to inspire visitors to protect the natural environment and wildlife. In 2017, Zhen Zhen gave birth to Xiang Xiang, her daughter, at the zoo.

日本东京上野动物园大熊猫馆：仿造四川的自然环境

日本东京上野动物园开放于 1882 年，占地面积约 14 公顷，展出动物约 420 种，大熊猫是这里最受欢迎的动物。

2011 年，来自中国大熊猫保护研究中心的大熊猫"比力"和"仙女"入住上野动物园大熊猫馆。通过向社会公开征名，园方将"比力"和"仙女"命名为"力力"和"真真"。

上野动物园的大熊猫馆"熊猫森林"仿照四川的树木、岩石、水流等自然景观建造，还原了大熊猫故乡四川的自然环境。馆内部分空间采用开放式设计，能激发游客们保护自然环境、保护野生动物的热情。2017 年，"真真"在上野动物园生下了女儿"香香"。

大熊猫"比力"和"仙女"旅居的上野动物园是日本第一座公共动物园。

The Ueno Zoo, where giant pandas Bi Li and Xian Nv sojourned, is Japan's first public zoo.

Giant Panda International Stars

大熊猫
国际明星

Ming: Adoration from Princess Elizabeth

In 1937, the months-old giant panda Ming was captured in Wenchuan by the Englishman Floyd Tangier Smith and sold to the London Zoo in the following year. Because of Ming, the zoo was inundated with numerous visitors. Even, the heir Princess Elizabeth, the future Queen of England, turned out to see Ming and became the Ming's fan. At the time of the World War Ⅱ, the arrival of Ming brought warmth and comfort to the British under the shadow of the war. As flames of war raged outside the zoo, Ming kept her composure, and her calm and optimism inspired the British people. Ming became such a spiritual icon that an anti-war documentary was made featuring the panda.

Ling Ling and Xing Xing: More than 8,000 People Braved the Rain to Greet Them at the Airport

In 1972, U.S. President Richard Nixon made his first visit to China, and the Chinese government decided to give two giant pandas as a gift to the American people. In April of the same year, the giant pandas Ling Ling and Xing Xing from Baoxing County in Sichuan Province took off from Beijing on a special airplane and arrived at the National Zoological Park in Washington, D.C. More than 8,000 people braved the rain to greet them at the airport. The American people called 1972 "The Year of the Giant Panda," and the two giant pandas frequently made headlines or appeared on the cover of major newspapers throughout the United States, earning them a reputation of World Animal Star. From then on, more than 3 million visitors came to the Washington National Zoo to see the giant pandas every year. Almost every family with children in the United States had panda toys because panda merchandise was a big seller.

"明"：女王储也是它粉丝

1937年，不足1岁的大熊猫"明"在汶川被英国人弗洛伊德·丹吉尔·史密斯捕获，第二年被卖给了伦敦动物园。因为"明"的加入，伦敦动物园一时间游客络绎不绝。连当时还不是英国女王的伊丽莎白女王储都去看望过"明"，并成了它的粉丝。时值二战，"明"的到来为在战争阴霾笼罩下的英国人带来了温暖的慰藉。外面的世界战火纷飞，而"明"在动物园里对一切泰然处之。它平和乐观的样子让英国人民备受鼓舞。"明"因此被树立为人们的精神偶像，人们甚至以它为主角拍摄了反战纪录片。

"玲玲"和"兴兴"：8000多人冒雨来接机

1972年，美国总统尼克松首次访华。中国政府决定将两只大熊猫作为国礼赠送给美国人民。同年4月，来自四川宝兴县的大熊猫"玲玲"和"兴兴"乘专机从北京起飞，到达位于华盛顿的美国国家动物园，8000多市民冒雨前来迎接。美国人民把1972年称为"大熊猫年"。"玲玲"和"兴兴"频频登上全美各大报刊的头版和封面，被称为"世界动物明星"。在这之后，每年有超过300万人次的游客来到华盛顿国家动物园参观大熊猫。在当时的美国，几乎每一个有孩子的家庭都有熊猫玩具，带有大熊猫图案的商品也十分畅销。

"玲玲"和"兴兴"
Ling Ling and Xing Xing

"明"
Ming

"梅梅"：它创造了三个奇迹

2000年，6岁的大熊猫"梅梅"为了中日大熊猫国际合作去到了日本。英雄母亲"梅梅"创造了总共生育9仔存活7仔的佳绩，孕育了海外最大的大熊猫家族——浜家族。2008年，"梅梅"因病抢救无效去世，享年14岁。2011年，日本和歌山政府向大熊猫"梅梅"授予了"县勋爵"称号。

"梅梅"在日本创造了三个奇迹：

它是首只秋季发情、冬季产仔的人工圈养大熊猫；

它一生共抚育成活7个孩子；

它母性很强，是全球首位自然哺育双胞胎宝宝成活的大熊猫妈妈。

"美兰"：安能辨我是雄雌

"美兰"2006年出生于美国亚特兰大动物园。它出生后被认定为雌性，经网络投票得名"美兰"，寓意"亚特兰大之美"。2010年，"美兰"回到家乡四川。成都大熊猫繁育研究基地为它举办了一场盛大的"择婿活动"。经网友投票，大熊猫"勇勇"成为"美兰"的配偶。然而，在它们共同生活一段时间后，"美兰"却出现了一系列雄性大熊猫性成熟期的反应，经检查，人们发现"美兰"原来是雄性。

2010年，"美兰"被选为世界自然基金会"地球1小时"全球推广大使。2011年3月26日，它在自己的圈舍里用嘴拉下吊绳，关闭灯光，让9个彩色冰雕熊猫停止融化，代表全球濒危动物在向人类吁请："保护不好黑白的我，世界将失去所有色彩。"

Mei Mei: Creator of Three Miracles

In 2000, 6-year-old giant panda Mei Mei traveled to Japan in the Sino-Japanese Giant Panda International Cooperation Program. The heroic mother gave birth to a total of 9 cubs, 7 of which survived. The largest overseas giant panda family, the Shirahama-born family in Japan, was bred. In 2008, Mei Mei died of illness at the age of 14 after failed rescue efforts. In 2011, the Wakayama government awarded the giant panda Mei Mei the title of Prefectural Lord.

Mei Mei has created three miracles in Japan:

She is the first captive giant panda to go into heat in the fall and give birth in the winter;

She has raised 7 children in her lifetime.

She is the first giant panda mother in the world to naturally feed twin babies without leaving either one of them dead.

Mei Lan: Can You Tell If I'm a Boy or a Girl?

Mei Lan was born at Zoo Atlanta in 2006. After birth, the giant panda was identified as female and named Mei Lan after an online vote, which means "the Beauty of Atlanta". In 2010, the giant panda returned to her hometown, Sichuan. The Chengdu Research Base of Giant Panda Breeding organized a grand "mate selection activity" for her. Giant panda Yong Yong was voted by netizens to be Mei Lan's mate. However, after they lived together for a period of time, Mei Lan showed a series of reactions of male giant pandas of sexual maturity. After examination, it was found that Mei Lan was actually a male.

In 2010, Mei Lan was selected as the global ambassador for WWF's Earth Hour. On March 26, 2011, he pulled down the lanyard with his mouth in his enclosure and turned off the lights to stop the 9 colorful ice pandas from melting, warning human beings on behalf of the globally endangered animals: "If you fail to protect the black and white species, the world will lose all its color."

"梅梅"和它的双胞胎宝宝
Mei Mei and her baby twins

"美兰"
Mei Lan

世界自然基金会
"地球1小时"全球推广
大使"美兰"
Mei Lan, the global ambassador for WWF's Earth Hour

联合国开发计划署
历史上首对动物形象大使
"启启"和"点点"。
The United Nations
Development Program
The first pair of animal
ambassadors in history, Qi Qi
and Dian Dian.

"启启"和"点点":双胞胎当上双大使

2015年9月,大熊猫"庆贺"在成都大熊猫繁育研究基地产下双胞胎宝宝。一个月后,它们被宣布成为联合国开发计划署历史上首对动物形象大使。

在为这两位形象大使开展的征名活动中,熊猫基地一共收到来自116个国家的热心网友提交的5000多个名字。在双胞胎大熊猫1岁生日之际,它们被正式命名为"启启"和"点点",代表2023年可持续发展议程的起点,也是全球发展新路径的起点。

Qi Qi and Dian Dian:
Twins Both Become Ambassadors

In September 2015, giant pandas Qing He gave birth to baby twins at the Chengdu Research Base of Giant Panda Breeding. A month later, they were announced as the first pair of animal ambassadors of UNDP in history.

The Panda Base received more than 5,000 names for the ambassadors from netizens of 116 countries. On their first birthday, the twin giant pandas were officially named Qi Qi and Dian Dian, which represent the starting point of the 2023 Agenda for Sustainable Development and a new path for global development.

"京京"和"四海":阿拉伯双星

2022年卡塔尔世界杯期间,来自中国大熊猫保护研究中心的大熊猫"京京"和"四海"在卡塔尔首都多哈的豪尔熊猫馆正式与公众见面。卡塔尔热烈欢迎中国大熊猫"京京"和"四海"的到来,并用阿拉伯文分别将它们命名为"苏海尔"和"索拉雅"。"苏海尔"是海湾地区可见的最亮恒星之一,"索拉雅"则是家喻户晓的昴宿星团的阿拉伯语名称。这是大熊猫首次旅居中东地区。

Jing Jing and Si Hai:
The Arabian Stars

During the 2022 FIFA World Cup in Qatar, giant pandas Jing Jing and Si Hai from the China Conservation and Research Center for Giant Panda officially met the public at the Al Khor Panda House in Doha, the capital of Qatar. They were warmly welcomed and gave them the Arabic names Suhail and Soraya. Suhail is the name of one of the brightest stars visible in the Gulf and Soraya is the Arabic name for the Pleiades, a well-known star cluster. This is the first time giant pandas have traveled to the Middle East.

"启启"和"点点"
Qi Qi and Dian Dian

2022年11月,"四海"和"京京"在卡塔尔多哈豪尔熊猫馆与游客见面。
In November 2022, Si Hai and Jing Jing met visitors at the Al Khor Panda House in Doha, Qatar.

Chapter 4　第四章

>> Human Impact
濒危年代

繁盛衰亡，跌宕起伏。
人与自然，和谐共生。

From the thriving to decline and fall, the history of the giant panda is full of peaks and valleys. Human beings and nature coexist harmoniously.

我们家族熬过了冰川纪大灭绝、具有超级进化技能、成年后几乎没有天敌……可是,如此彪悍的我们怎么熬着熬着就把自己熬成了珍稀动物呢?除了自然环境的影响,人类活动的影响也是重要因素!

你知道吗,每只成年大熊猫大概需要5平方千米的领地(相当于大概700个标准足球场大小)才能保证生存!可是,随着历史前行,人口持续增长,人类发展所需的资源越来越多。人类的活动范围不断扩张,我们的家园被挤占,族群被迫分隔,栖息地支离破碎,种群数量大幅下降。大自然是一个相互依存、相互影响的生态系统。我们的生存环境被破坏,受到伤害的不仅仅是我们大熊猫一族,也包括和我们生活在一起的动植物邻居们。如果我们陷于困境,跟我们生活在同一个地球家园的人类难道能独善其身吗?

>> Our family survived the Ice Age extinction and mastered super evolutionary skills, leaving us with no natural enemies after entering into adulthood. But, how did such species end up being a rare animal? In addition to the influence of natural environment, human activities are also a significant factor.

Did you know that each adult giant panda needs a territory of about 5 square kilometers (equivalent to about 700 standard soccer fields) to survive? However, as human populations continue to soar with each passing year, they call for an increasing number of resources for their development and expand their scope of activities, occupying our homes, separating our communities, and fragmenting our habitats. As a result, our populations have been nose-diving. Nature is an interdependent ecosystem. The destruction of our living environment will not only deal a heave blow to us, but also to our neighbors. If we are at risk, can human beings who live on the same planet with us be immune from the danger?

嗨,我是小川。
Hi, I'm Xiao Chuan.

\>\>

Chengdu
Giant Panda
Museum
Exhibition Hall 4
Human Impact

成都大熊猫博物馆 第四展厅 濒危年代

Wild Home for Giant Pandas

大熊猫的野外家园

大熊猫栖息地
>> Giant Panda Habitat

Giant pandas inhabit subalpine regions at an altitude of 1,500 to 3,500 meters. In these warm and humid areas with a temperature below 20℃ all year round, mountains roll and rivers flow. These areas are also rich in forest resources, especially a wide variety of bamboo.

Historically, giant pandas were distributed in many areas. However, as the economy and population grew, mankind have exploited various resources. As a result, the distribution areas of giant pandas have narrowed westward. Now, giant pandas can only be seen in the Qinling, Min, Daxiangling, Xiaoxiangling, Qionglai and Liang Mountains across Sichuan, Shaanxi, and Gansu Provinces.

With a rich variety of species, Sichuan Giant Panda Sanctuaries are one of 36 global biodiversity hotspots and are home to nearly 75% of wild giant pandas. It is also home to red pandas, golden snub-nosed monkeys, snow leopards, and other rare wild animals.

大熊猫一般栖息在海拔 1500～3500 米的亚高山地区，那里山峦起伏、溪流纵横、森林资源非常丰富，竹子种类繁多，气候温润潮湿，气温常年低于 20℃。

历史上，大熊猫分布区域广泛，但随着经济的发展和人口数量的增长，人类不断向自然界索取各种资源，致使大熊猫的分布区向西退缩。目前，大熊猫仅分布在四川、陕西、甘肃三省的秦岭山系、岷山山系、大相岭山系、小相岭山系、邛崃山系、凉山山系这六大狭长地带。

四川大熊猫栖息地拥有丰富的物种种类，是全球 36 个生物多样性热点地区之一，近 75% 的野生大熊猫栖息于此。除了大熊猫，这里也是小熊猫、川金丝猴、雪豹等珍稀野生动物栖息的地方。

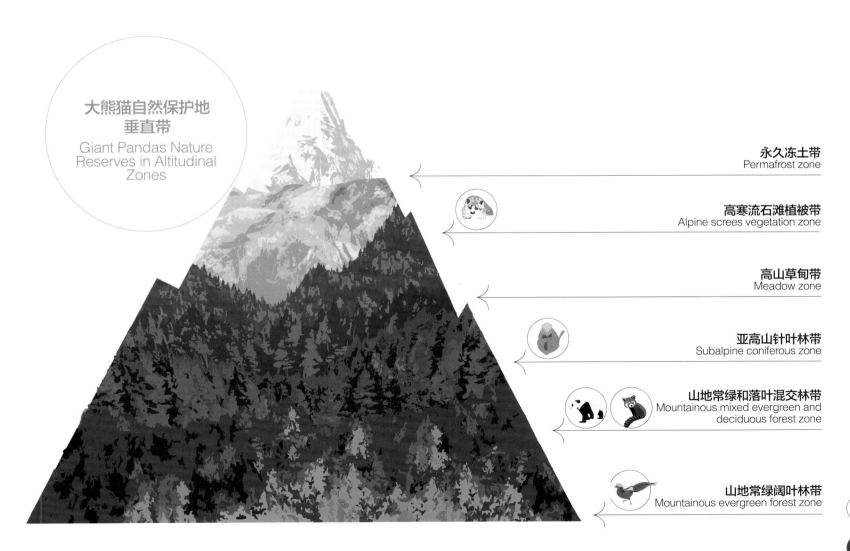

大熊猫自然保护地垂直带
Giant Pandas Nature Reserves in Altitudinal Zones

永久冻土带
Permafrost zone

高寒流石滩植被带
Alpine screes vegetation zone

高山草甸带
Meadow zone

亚高山针叶林带
Subalpine coniferous zone

山地常绿和落叶混交林带
Mountainous mixed evergreen and deciduous forest zone

山地常绿阔叶林带
Mountainous evergreen forest zone

大熊猫的野外栖息地植被茂密,气候温润潮湿。
Their wild habitat is characterized by dense vegetation and a temperate, humid climate.

Factors Concerning Giant Pandas' Habitat Selection

大熊猫栖息地选择的要素

食物：竹林繁茂但不能过密
>> Food: Luxuriant but Sparse Bamboo Forest

Giant pandas prefer one food-bamboo. So, when giant pandas choose their habitats, they will first consider whether the bamboo grows plentifully. Bamboos vary from season to season and distribution areas, and more than 20 species of bamboo are food to giant pandas. Research suggests that giant pandas do not prefer one species. They eat all species of bamboo in their habitats.

There are usually 3 to 4 species of bamboo in one mountain system, and giant pandas choose the area with various and thick bamboo as their habitats. However, giant pandas are seldom seen in areas with dense bamboo forests, because it is difficult for them to acquire their food in dense bamboo forest. Besides, the bamboo is not as good in dense forest. A clearing usually has thick bamboo because other trees are cut down. However, giant pandas will not live in a clearing because there are no tall trees and rich biodiversity.

大熊猫以竹子为食，食物种类较为单一，所以，竹类的生长情况是大熊猫选择栖息地的首要因素。不同品种竹子的生长季节、分布范围不同，其中被大熊猫作为主食的竹子有 20 多种。研究发现，大熊猫并没有特别偏爱的竹子品种，生长在它们生活区域内的各种竹子都可以成为它们的食物。

每个山系通常只生长有 3~4 种竹子。大熊猫会选择那些竹子品种多、竹林繁茂的地方作为自己的栖息地。但是，竹林生长过于繁密的地方却很少能发现大熊猫的足迹。这是因为过密的竹林不利于大熊猫穿行，会给它们取食带来一定的难度；同时，竹林过密，也会导致竹子的品质下降；而且，竹子生长得特别繁茂的地方多是林木采伐后的空地，这里的上层空间没有高大的乔木覆盖，生物多样性比较单一，所以大熊猫不会选择在这样的地方生活。

大熊猫最爱吃这些竹子
Bamboo Giant Pandas Like

拐棍竹
Fargesia robusta

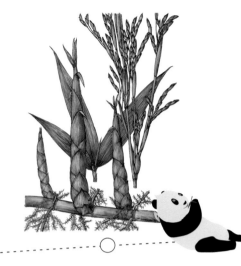

苦竹
Pleioblastus amarus

冷箭竹
Arundinaria faberi

拐棍竹

禾本科 箭竹属

该种竹类为邛崃山系和岷山山系大熊猫的主食竹，生长海拔1300~1900米。拐棍竹喜阴，在阴暗环境下能生长到5米高，每年出笋两次。野生大熊猫采食竹叶和竹竿，圈养大熊猫主要在春夏季喂食该种竹类。

苦竹

禾本科 苦竹属

分布在长江流域、河南山区，生长在海拔600米左右的地带。苦竹属于低山平坝竹，春季出笋，圈养大熊猫喜食当年生的竹竿，是成都大熊猫繁育研究基地大熊猫冬季的主要后备口粮。苦竹营养价值较低。

冷箭竹

禾本科 巴山木竹属

分布于四川西部和贵州部分地区，生长海拔2300~3500米。该品种为四川卧龙、宝兴、天全、马边大风顶等自然保护区大熊猫的主食竹类。该竹类于20世纪80年代大面积开花，导致许多野生大熊猫饿死。

Fargesia robusta

Gramineae *Fargesia*

This species is the food of giant pandas in the Qionglai and Min Mountains. It grows at an altitude of 1,300 to 1,900 meters. This shade-loving bamboo can be as high as 5 meters in a dark environment. The bamboo shoots appear twice a year. Wild giant pandas eat the leaves and stems, and captive giant pandas are fed this species of bamboo in spring and summer.

Pleioblastus amarus

Gramineae *Pleioblastus*

This species grows particularly in flat and low mountains at an altitude of about 600 meters in the Yangtze River basin and the mountainous regions of Henan province. The bamboo shoots appear in spring. Captive giant pandas love the bamboo stem of this year. It is the main reserved food for giant pandas in the Chengdu Research Base of Giant Panda Breeding in winter. It has little nutrition.

Arundinaria faberi

Gramineae *Bashania*

This species grows at an altitude of 2,300 to 3,500 meters in western Sichuan province and part of Guizhou province. It is consumed by the giant pandas in Wolong, Baoxing, Tianquan, and Mabian Dafengding Nature Reserves of Sichuan Province. The species of bamboo bloomed extensively and withered in the 1980s, causing many wild giant pandas to starve to death.

巴山木竹
Arundinaria fargesii

白夹竹
Phyllostachys bissetii

箬竹
Indocalamus tessellatus

Arundinaria fargesii
Gramineae *Bashania*

This species grows at an altitude of 1,100 to 2,500 meters adjacent to the Daba, Micang, and Qinling Mountains from eastern to northern Sichuan Province, western Hubei Province, southern Shaanxi Province, and southeastern Gansu Province. This species of bamboo has high nutritional value and good palatability, and is the staple food of giant pandas in the Qinling Mountain and those in the Chengdu Research Base of Giant Panda Breeding.

Phyllostachys bissetii
Gramineae *Phyllostachys*

This species grows in Sichuan Province and Zhejiang Province, particularly in flat and low mountains at an altitude below 1,000 meters. The bamboo shoots appear in May and are the food of giant pandas in late spring and early summer. The bamboo stems have little nutritional value. Captive giant pandas peel off the stems and eat them in spring.

Indocalamus tessellatus
Gramineae *Indocalamus tessellatus*

The species grows in southern China, particularly in the vast river basin south of the middle and lower reaches of the Yangtze River, and is found in low valleys below 1500 meters. Because of the large leaves of the bamboo, giant pandas only feed on its leaves. In late winter, the leaves go bad because of the frost and snow and giant pandas eat less because of it.

巴山木竹

禾本科 巴山木竹属

分布于四川东部至北部、湖北西部、陕西南部和甘肃东南部的大巴山、米仓山、秦岭相毗连的地区，生长海拔 1100～2500 米。该种竹类营养价值较高，适口性良好，为秦岭大熊猫的主食竹类，也是成都大熊猫繁育研究基地大熊猫的主食之一。

白夹竹

禾本科 刚竹属

主要分布于四川、浙江等省，属于低山平坝竹，生长在海拔 1000 米以下。白夹竹每年 5 月出笋，其竹笋是大熊猫晚春初夏的主要食物之一。白夹竹竹竿的营养价值较低。圈养大熊猫主要在春季采食该种竹类的竹竿，采食时会将外面坚硬的皮剥掉。

箬竹

禾本科 箬竹属

普遍见于南方地区，尤其是长江中下游以南广大流域，生长在海拔 1500 米以下的低山谷间等地。箬竹竹叶较大，大熊猫只采食其竹叶。深冬由于霜雪影响，竹叶质量变差，大熊猫的食用量也会减少。

白夹竹
Phyllostachys bissetii

水源：近水处最宜居
>> Water: Living Next to Water

饮水对大熊猫来说与食竹同样重要。竹林茂盛、植被覆盖良好的地方水源通常都比较丰富。大熊猫一般不饮用静水，也不以冰雪补充水分，而是饮用河水或泉水。

Water is as important as bamboo for giant pandas. Places with lush bamboo forests and vegetation are usually rich in water. Giant pandas do not drink still water nor ice or snow and instead directly drink from rivers and springs.

Trees: Raising Cubs in Tree Dens

大树：树洞是育儿的好地方

Giant pandas inhabit coniferous forests, broad-leaved forests, and mixed coniferous and broad-leaved forests. Tall trees are important for giant pandas due to allowing to rest and avoid their enemies. Patrol officers in sanctuaries often see giant pandas napping and resting on trees in the afternoon. The dens at the foot of these trees are the best place for giant pandas to give birth and raise their babies.

大熊猫栖息的地方，为针叶林、阔叶林、针阔叶混交林地带。森林中的大树挺拔繁茂，为大熊猫提供了休息和躲避天敌的重要场所。大熊猫栖息地巡护人员经常能在午后看到大熊猫在大树上打盹休息。苍劲挺拔的古树基部的树洞是大熊猫产仔育幼的上选场所。

海拔：上下迁移就为吃
Altitude: Migrating for Food

海拔高度是大熊猫栖息地选择最直接的条件。因为海拔高度与植被、温度、水源等自然条件都紧密相关。大熊猫为了常年能吃到最可口、最有营养的食物，会随季节变化在不同海拔地带和植被类型中来回迁移。比如秦岭山系，在海拔 1900～3100 米的地带生长有秦岭箭竹，这是夏季大熊猫的主要食物来源。等到了冬季，大熊猫就会向山下移动，去采食在海拔 1100～1900 米地带生长着的巴山木竹。

Altitude is the most relevant condition for determining giant panda habitats because it is closely related to vegetation, temperature, water, and other natural conditions. Giant pandas migrate between different elevations and vegetation depending on seasons for tasty and nutritious food. For example, giant pandas mainly eat *Fargesia qinlingensis* in the summer at an altitude of 1,900 to 3,100 meters in the Qinling Mountains. In the winter, giant pandas move downhill for *Arundinaria fargesii* at an altitude of 1,100 to 1,900 meters.

坡度：平地和缓坡更省力
>> Slope: Saving Energy on Flat and Gentle Slopes

Giant pandas usually inhabit flat and gentle slopes, which not only helps them forage, but also does not consume much precious energy to climb. Wild research has revealed that giant pandas prefer places with slopes below 30 degrees, and rarely visit areas with slopes above 40 degrees.

大熊猫选择的栖息地一般都比较平坦，坡度不大，这样既有利于觅食，也不会因为耗费太多体力爬坡而消耗宝贵的能量。野外研究证实，大熊猫喜欢在坡度 30 度以下的地方活动，很少光临坡度超过 40 度的地方。

Survey on Wild Giant Pandas
大熊猫野生资源调查

为了了解大熊猫的种群数量、栖息地范围以及栖息地内的其他野生动植物资源情况，大约每隔10年，中国政府就会组织进行一次大熊猫野生资源调查，每次调查耗时3~5年。

1974~1977年、1985~1988年、1999~2003年和2011~2013年，中国政府共组织了四次大熊猫野生资源调查，历次调查取得的成果为有效保护大熊猫提供了科学依据。

In order to have a clear picture of their population size, the extent of their habitat, and other wildlife resources in their habitat, the Chinese government organizes a wildlife resources survey of giant pandas about every decade, which lasts 3 to 5 years.

1974 ~ 1977, 1985 ~ 1988, 1999 ~ 2003 and 2011 ~ 2013, four surveys were carried out, whose results have provided a science-based basis for the effective conservation of giant pandas.

1974 ～ 1977

第一次调查
>> First Survey

The survey team with 18 members utilized a route survey and concluded there were nearly 2,459 wild giant pandas. Although the data was inaccurate due to the simplicity of the survey tools, it has given people an idea of the approximate habitat range of the giant pandas. After the survey, 3 giant panda conservation observatories were established in Sichuan Province and Gansu Province, including the Wuyi Camp Observatory in Wolong, Sichuan Province, which is still active now.

第一次大熊猫野生资源调查的调查队由 18 名队员组成，采用"路线调查法"，得出野生大熊猫的数量大约为 2459 只。虽然因为调查工具简单，导致调查数据并不准确，但是人们由此掌握了大熊猫大致的栖息地范围。随着这一次调查，四川、甘肃两省先后设立了三个大熊猫保护观测站，其中就包括保留至今的四川卧龙"五一棚"观测站。

2459 只

第一次大熊猫野生资源调查
First Survey on Wild Giant Pandas

第二次调查 >> Second Survey

1985 ～ 1988

第二次调查的调查队由 30 名队员组成，采用"拉网式路线调查法"，主要根据采集到的大熊猫粪便分析大熊猫数量。时值岷山、邛崃山的箭竹大面积开花死亡，大熊猫处于"饥荒"之中。调查显示，全国野生大熊猫约为 1114 只，其栖息地总面积约 1.39 万平方千米。

The 30-member survey team utilized a dragnet route survey and analyzed the number of giant pandas based on the collected feces. During the survey, Fargesia spathacea in the Min and Qionglai Mountains blossomed and died, and giant pandas were experiencing a famine. The survey revealed there were about 1,114 wild giant pandas with a total habitat area of around 13,900 square kilometers across China.

第二次大熊猫野生资源调查
Second Survey on Wild Giant Pandas

1114 只

1999 ~ 2003

第三次调查
>> Third Survey

This time, the survey team collected and analyzed 3,800 giant pandas' feces. The team estimated that there were about 1,596 wild giant pandas in a total habitat area of around 23,000 square kilometers based on the distance and bamboo stem segments.

这次，调查队采用了"距离区分法"和"咬节分析法"进行调查，总共收集分析了 3800 份熊猫粪便，经过研究推算出野生大熊猫数量约为 1596 只，大熊猫栖息地总面积约为 2.3 万平方千米。

第三次大熊猫野生资源调查
Third Survey on Wild Giant Pandas

1596 只

第四次调查 / Fourth Survey

2011 ~ 2013

在沿用第三次的调查方法对大熊猫种群数量进行统计的基础上，调查队还引入了非损伤性 DNA 检测技术对野生大熊猫的遗传多样性进行分析，同时利用遥感卫星数据对大熊猫栖息地状况进行评估。调查结果显示，截至 2013 年底，全国野生大熊猫的数量约为 1864 只，其中四川省 1387 只，陕西省 345 只，甘肃省 132 只。大熊猫栖息地总面积约 258 万公顷，潜在栖息地面积 91 万公顷。四川省在大熊猫种群数量和种群密度方面都排名第一。

In addition to the statistics of the giant panda population using the methods utilized in the third survey, the survey team also introduced non-damaging DNA detection technology to analyze the genetic diversity of wild giant pandas, and assessed the habitat condition of giant pandas using data from remote sensing satellites. The results showed that by the end of 2013, there were 1,864 giant pandas, including 1,387 in Sichuan Province, 345 in Shaanxi Province and 132 in Gansu Province. The total habitat area was about 2.58 million hectares and a potential habitat area was 910,000 hectares. Sichuan Province ranked first in giant panda population quantity and density.

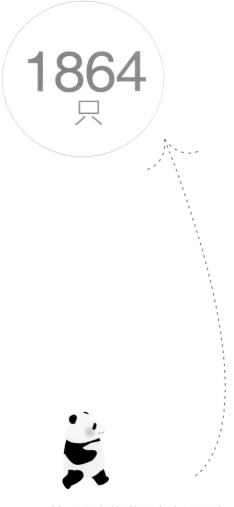

1864 只

第四次大熊猫野生资源调查
Fourth Survey on Wild Giant Pandas

大熊猫野生资源调查的"神秘武器"
Useful Devices in the Survey on Wild Giant Pandas

相比老一辈科研人员只能凭借罗盘、高程图这些设备来进行野外调查，现在的科研人员在开展大熊猫野生资源调查时有了更科学的研究技术和更先进的调查工具。

Compared to previous researchers that conducted field surveys with compass and elevation maps, and current researchers now conduct surveys on wild giant pandas with more scientific technologies and more advanced devices.

01

红外自动触发相机
Infrared-triggered camera-trapping

利用红外感应设备，在无人操作的情况下，持续性、长期性地自动捕捉野生动物的活动影像，可用于开展动物多样性、动物活动节律、动物空间分布等科学研究。

The infrared sensing equipment can be used to continuously capture the activity of wild animals without requiring staff to be present. The equipment can be used to conduct scientific research on animal diversity, animal activity rhythms, and animal spatial distributions.

02

高精度照相器材
Highly-precise photographic apparatus

用于记录大熊猫在野外活动的珍贵影像资料。

The apparatus can be used to record precious images and videos of giant pandas' activities in the wild.

03

游标卡尺、皮尺、卷尺等
Vernier caliper, tape measure, etc

配合使用，用于大熊猫等动物的栖息地调查，测量调查样方大小、可食竹的高度和基径、乔木胸径、大熊猫粪便的体积和咬节等数据。

These tools are used to survey the habitats of giant pandas and other animals, including the size of quadrat, the height and branch diameter of edible bamboo, diameter of trees at breast height, volume of giant panda feces and bamboo stem segments.

04

无人机

远距离监测大熊猫等野生动物的活动情况，具有方便、快捷、不易打扰动物等优点。

Drone

Drones can monitor the activities of giant pandas and other wild animals from a distance. It is convenient and fast all while not disturbing the animals.

05

无线电 GPS 项圈

无线电颈圈加载 GPS 功能，将其佩戴于大熊猫颈部，可帮助科研人员远距离确定大熊猫的位置，代替人工实时记录动物的行动轨迹，有助于研究大熊猫的活动范围、活动节律等。

Radio GPS collars

The radio collars equipped with GPS functionality can help researchers determine the position of the giant panda from a distance, replacing manual real-time recording of the animal's trajectory, and help study the activity range and rhythm of the giant panda.

06

手持 GPS

研究人员在野外工作期间，可以将其用于导航、记录巡护航迹、记录动物粪便发现点。

Hand-held GPS

During their field work, researchers can use it for navigating their routes to record patrol trails and document animal fecal finds.

Survival Crisis Facing Giant Pandas

大熊猫的生存危机

Compared with the third survey, the fourth one reveals a steadily increasing number of wild populations of giant pandas, significantly expanding habitat, rapidly growing captive populations, as well as better conservation and management capabilities. However, giant panda conservation still has a long way to go as a large number of factors still exist, which threaten the survival and reproduction of giant pandas. Since ancient times, natural disasters have been the main factor affecting the growth of the giant panda population. But with humans' insatiable demand for natural resources, nowadays "man-made calamities" turn out to be more life-threatening than natural disasters.

与第三次大熊猫野生资源调查相比，第四次野生资源调查显示大熊猫野生种群数量稳定增长，栖息地范围明显扩大，圈养种群规模快速发展，保护管理能力逐步增强。但是，大熊猫保护的形势依然十分严峻，大量威胁大熊猫种群生存、繁衍的因素依然存在。从古至今，自然灾害一直是影响大熊猫种群发展的主要因素，但随着人类对自然资源无节制地索取，如今"人祸"更胜于"天灾"。

Influence of the Natural Environment

Climate Change: Disappearance of a Third of Bamboo Forests in the Future

In recent years, global climate change, such as rising temperatures and erratic precipitation, has put many wild animals in a difficult situation. Some reports say that climate change could reduce the area of bamboo forests that giant pandas rely on by more than 35% over the next 80 years. If that happens, giant pandas will be confronted with a food shortage and other crises. It will exert an immeasurable impact on the survival of giant pandas.

Bamboo Blossoming: Food Shortages in the "Isolated Island" Dilemma

Bamboo, the giant panda's staple food, blossoms largely every time it goes through a certain cycle. Bamboo withers after blooming, which will put giant pandas at risk of dying from food shortage. From 1974 to 1976, the investigation team of the State Forestry Administration of the People's Republic of China (now the National Forestry and Grassland Administration) found 138 dead giant pandas in withered bamboo forests. Post autopsy, researchers found there was no food in their stomachs. In 1983, the Fargesia spathacea in the Qionglai Mountains largely blossomed and withered, and more than 500 giant pandas were at risk of starving.

In giant panda habitats, bamboo blossoming should not have threatened the survival of giant pandas. When one species of bamboo blooms and withers, giant pandas can migrate elsewhere in search of other edible bamboo. However, due to years of population growth, industrial development, urban sprawl, and other reasons, the giant panda's habitat has been severely damaged and separated into "isolated islands". As a result, giant pandas are unable to migrate long distances in search of other species of bamboo, leaving them struggling for food.

自然环境的影响

气候变化：未来竹林可能减少三分之一

近些年，气温上升、降水无常等全球性的气候变化，使许多野生动物陷入了生存困境。有报告指出，气候变化可能在未来80年中造成大熊猫赖以生存的竹林面积减少35%以上。如果这种情况真的发生，野生大熊猫将面临食物短缺等危机，这将对大熊猫的生存产生不可估量的影响。

竹子开花："孤岛"困境导致食物紧缺

大熊猫的主食竹子每经历一定周期就会大面积开花。竹子开花后会枯萎，导致大熊猫因缺少食物而死亡。1974年至1976年，国家林业局调查队在开花后枯萎的竹林中，发现了138具大熊猫尸体。通过解剖，人们发现这些大熊猫的胃里没有食物。1983年，邛崃山系的箭竹大面积开花枯萎，500多只大熊猫也曾面临饿死的危险。

其实，在大熊猫的栖息范围内，竹子开花本不应该威胁到大熊猫的生存，因为当一种竹子开花枯死后，大熊猫可以迁徙到别处去寻找其他可食竹类。但是，由于多年来人口增长、工业发展、城镇扩张等原因，大熊猫的栖息地遭到了严重破坏，被分隔成了相互隔离的"孤岛"，这致使大熊猫无法远距离迁移去寻找其他种类的竹子，从而陷入缺少食物的困境。

竹子开花
Bamboo blossoming

人类活动的影响
Influence of Human Activities

栖息地破碎化：无奈的近亲繁殖

大熊猫栖息地多为山区，"靠山吃山"是当地居民世代沿袭的生存之道。多年来，随着人口增长和生活方式的改变，人们对自然资源的索求日益增加。他们进山挖笋、放牧、开荒、采药、砍伐，甚至进行非法捕猎，这些行为严重威胁着栖息地内大熊猫种群的生存，迫使它们背井离乡，流离失所。由于自然隔离和人为干扰等因素的影响，大熊猫野外种群被分割成33个局域种群，目前有24个局域种群具有较高的生存风险。栖息地的丧失和破碎化，给大熊猫的种群交流造成了严重阻碍，导致了严重的近亲繁殖现象，使大熊猫后代个体生存与环境适应能力显著下降。处于相对隔离状态的小群体中，容易产生基因频率的随机波动，若不加以改善，物种将逐渐消失，而这正是大熊猫种群所面临的现状。

生态环境恶化：威胁无处不在

随着社会经济的发展，城市人口剧增，人类活动对环境产生了巨大的破坏性影响。例如，大肆架桥修路，兴修水电站，过度开发旅游度假区，导致森林调节功能下降，水土流失严重；使用一次性塑料制品，塑料制品难以被降解，成为许多动物的杀手；垃圾焚烧产生有害气体，造成大气污染；工业排污、汽车尾气排放，导致温室效应严重，大气变暖，海平面上升……生态环境的恶化会进一步影响生态系统的稳定，这对大熊猫和其他野生动物来说都是巨大的威胁。

Fragmented Habitats: Inevitable Inbreeding

Giant pandas live mostly in mountainous areas, where people have lived for generations. However, in the midst of population growth and lifestyle changes, mankind demands more natural resources. They go into the mountains to dig bamboo shoots, graze animals, clear land, collect medicine, cut down vegetation, and even poach. These behaviors have posed a serious threat to the survival of the giant panda population in the mountains and made giant pandas homeless and displaced. Influenced by natural isolation and mankind, the wild giant panda population has been divided into 33 local populations. At present, 24 local populations are at high risk of survival. The loss and fragmentation of habitat has seriously hindered the population exchange of giant pandas, and led to serious inbreeding, which has significantly reduced the individual survival and environmental adaptability of giant panda offspring. In small populations in relative isolation, there are random fluctuations in gene frequency, and if not improved, the species will gradually disappear. This is exactly what the giant panda population is facing.

Deteriorated Ecological Environment: Full of Threats

With the development of social economy, the population in cities has increased dramatically, and their activities have had a huge destructive impact on the environment. For example, the large-scale construction of bridges and roads, and hydropower stations, and the over-development of tourist resorts have led to the decline of forest regulation functions and serious soil erosion; the use of single-use plastic products which do not degrade has become a killer of many animals; garbage incineration produces harmful gases and causes air pollution; industrial emissions and automobile exhaust have led to serious greenhouse effects, atmospheric warming, and sea level rise... The deterioration of the ecological environment will have a negative impact on the stability of the ecosystem, which is a huge threat to giant pandas and other wild animals.

和大熊猫面临同样困境的中国珍稀保护动物

>> Rare Protected Animals in China Facing the Same Plight as Giant Pandas

麋鹿
David's deer

朱鹮
Crested ibis

川金丝猴
Golden snub-nosed monkey

褐马鸡
Brown-eared pheasant

小熊猫
Red panda

华南虎
South China tiger

黑颈鹤
Black-necked crane

扬子鳄
Chinese alligator

白鳍豚
White-flag dolphin

华北豹
North China leopard

藏羚羊
Tibetan antelope

Chapter 5　第五章

>> Helping Hands
保护之路

科研保护，技术创新。
人类护佑，生生不息。

Protected by experts, giant pandas grow strong. Technological innovation empowers conservation through scientific research.

进入20世纪后，随着人类世界工业化的快速发展，大自然的生态遭到了严重破坏，我们大熊猫家族面临着前所未有的生存危机。值得庆幸的是，人类已经认识到了保护生态环境的重要性。从20世纪50年代起，一代代"熊猫人"全力以赴地投入到对我们家族的保护工作中。

"太阳出来啰喂，照亮我也照亮你，一样的空气我们呼吸，这世界，我和你生活在一起。请让我来帮助你，就像帮助我自己；请让我去关心你，就像关心我们自己。这世界，会变得更美丽……"1983年夏季，岷山和邛崃山系的高山箭竹大面积开花枯萎，我们家族面临粮食危机。这首名叫《熊猫咪咪》的歌曲应运而生，打动了国内外无数热爱大熊猫、关注自然生态的人们，"拯救大熊猫"的热潮在全世界掀起。就如歌中所唱，保护以我们为代表的珍稀物种，保护生物多样性，就是保护整个地球家园。

>> Since the beginning of 20th century, booming industrialization in the human world has wrecked havoc on the ecology, posing an unprecedented existential crisis to us. Thankfully, humans have recognized the importance of protecting the ecological environment. Since the 1950s, generations of people devoted to conserving giant pandas have stayed committed to the protection of our population.

"The sun comes out and shines on me and you; we breathe the same air and live in the same world; please let me help you as I help myself, and please let me care for you as we care for ourselves. This world will become more beautiful..." In the summer of 1983, *Fargesia spathacea* in the Min and Qionglai Mountains bloomed and withered en masse, inflicting a food crisis on us. The first few lines come from a song called *Panda Mimi* which was composed in response to the disaster. The song moved countless people at home and abroad who love giant pandas and are concerned about the ecology, starting the craze for saving giant pandas across the globe. As the song goes, to protect the rare species we represent and conserve biodiversity is like safeguarding the entire planet.

嗨，我是小川。
Hi, I'm Xiao Chuan.

>>

Chengdu
Giant Panda
Museum
Exhibition Hall 5
Helping Hands

成都大熊猫博物馆 第五展厅 保护之路

Laws Underpinning Giant Panda Conservation

大熊猫保护的法制底气

大熊猫保护的法制建设
Legal Construction of Giant Panda Conservation

After Armand David introduced giant pandas to the world in 1869, three "giant panda craze" have been set off around the world. Before the founding of the People's Republic of China, the government then was weak, wars were ongoing and people did not realize the importance of wild animal conservation. From 1870 to 1949, giant pandas were hunted and captured by westerners, which greatly damaged the giant panda population.

After the People's Republic of China was founded, the government has strengthened the conservation of giant pandas and other rare wild animals. The launch of the *Measures for Rare Organisms Protection* in 1950 marked the beginning of the legal construction of wild animal conservation in China.

In 1962, the State Council issued the *Instruction on the Active Protection and Rational Use of Wildlife Resources*, which listed giant pandas and other rare species as hunting-prohibited animals. In 1988, the government issued the *Law of the People's Republic of China on the Protection of Wildlife*. This law not only includes giant pandas as a state first-class key protection wildlife, but also provides habitat conservation of giant pandas and other rare wildlife and serious punishments for violations, such as smuggling, capturing, and killing rare wild animals. By doing so, the conservation of giant pandas and other rare wild animals has truly achieved a legal basis.

After that, the State Council and the Ministry of Forestry of the People's Republic of China (now the National Forestry and Grassland Administration) have formulated and issued a batch of administrative and ministerial laws and regulations. In addition, the *Law of the People's Republic of China on Environmental Protection* and *Law of the People's Republic of China on the Control of Firearms* also has corresponding provisions on the protection and management of rare animals such as giant pandas.

Now, the legal system of wildlife conservation in China has been established and improved. More and more giant panda habitats are being designated as Nature Reserves, and giant panda national parks have been established. The public has been becoming more aware of the conservation of giant pandas and other wild animals. In China, giant pandas and other wild animals have been increasingly well conserved.

1869 年，阿尔芒·戴维将大熊猫介绍给全世界后，世界范围内掀起了三次"大熊猫热"。中华人民共和国成立前，由于政府软弱无力，加上战争不断，以及人们野生动物保护意识的缺乏，在 1870 年到 1949 年间，大熊猫不断遭到西方人的猎杀和捕捉，对大熊猫种群造成了极大破坏。

中华人民共和国成立后，政府加强了对大熊猫等珍稀野生动物的保护管理，于 1950 年颁布了《稀有生物保护办法》，中国野生动物保护的法制建设从此起步。

1962 年，国务院发布了《关于积极保护和合理利用野生动物资源的指示》，大熊猫等珍稀物种被列为国家禁猎动物。1988 年，政府颁布了《中华人民共和国野生动物保护法》，不仅明确地将大熊猫列为"国家一级重点保护野生动物"，还就大熊猫等珍稀野生动物的栖息地保护，走私、捕杀珍稀野生动物等违法行为将受到的严厉惩罚等做出了明确规定，使大熊猫等珍稀野生动物的保护工作真正做到了有法可依。

这之后，国务院和林业部又先后制定并发布了一批行政法规和部门规章。此外，《中华人民共和国环境保护法》《中华人民共和国枪支管理法》等法律也都对大熊猫等珍稀动物的保护、管理作了相应规定。

现在，中国野生动物保护的法律体系已经基本确立并在不断完善，越来越多的大熊猫栖息地被划建为自然保护区，还建立起了大熊猫国家公园，公众保护大熊猫、保护野生动物的意识也越来越强，大熊猫等珍稀野生动物受到了越来越完善的保护。

大熊猫的野外栖息地
Wild Habitat of Giant Pandas

Two Methods of Conservation

两种保护方式

In-situ Conservation: Most Direct and Effective

What is *in-situ* conservation? It is the main form of species conservation that is meant to conserve the species without moving them out of their habitats. A giant panda Nature Reserve is the most direct and effective way of conservation. It controls human activities outside the reserve and minimizes the impact of various factors on the giant panda. In order to protect the giant panda and its habitat, 67 giant panda nature reserves have been established across the country, forming a giant panda habitat protection network system, and the endangered situation of the giant panda in the wild has been further alleviated.

67 Nature Reserves: Conserving nearly 70% of Wild Giant Pandas

The government has set aside certain areas in the wild for special conservation and management to protect giant pandas, which are called the giant panda Nature Reserve. The Nature Reserve protects not only the giant pandas, but more importantly, the habitat that the giant panda depends on, and many companion species of the giant panda benefit. The first giant panda Nature Reserve was established in 1963. By the end of the fourth national survey on wild giant pandas, China has established 67 giant panda Nature Reserves, with a total area of 3.36 million hectares, effectively conserving 53.8% of giant panda habitats and 66.8% of the wild giant panda population.

什么是就地保护？就地保护就是指不将物种迁移出其原来的栖息区域而就地实行保护，这是物种保护的主要形式。建立大熊猫自然保护区，将人为干扰控制在保护区外，尽可能将各种因素对大熊猫的影响降至最低，是保护大熊猫最直接有效的途径。为保护大熊猫及其栖息地，全国已建立了 67 个大熊猫自然保护区，形成了大熊猫栖息地保护网络体系，野外大熊猫的濒危状况得到了进一步缓解。

67 个自然保护区：守护近七成野生大熊猫

为了保护大熊猫，政府在野外大熊猫分布的区域内划出一定面积予以特殊保护和管理，这些区域就是大熊猫自然保护区。大熊猫自然保护区的作用不仅仅是保护大熊猫这一个物种，更重要的是保护了大熊猫赖以生存的栖息地和大熊猫的同域伴生物种。从 1963 年中国最早的大熊猫保护区成立，到全国第四次大熊猫野生资源调查结束，全国有大熊猫分布的自然保护区达 67 个，总面积达 336 万公顷，53.8% 的大熊猫栖息地和 66.8% 的野生大熊猫种群被纳入自然保护区的有效保护中。

大熊猫国家公园：圈进七成大熊猫栖息地

2017年，国家启动了大熊猫国家公园建设。大熊猫国家公园总面积2.20万平方千米，是全球生物多样性保护的热点地区之一，也是大熊猫野生种群的核心分布区。野生大熊猫数量以岷山山系最多，邛崃山山系次之，大相岭、小相岭和秦岭山系数量较少。大熊猫国家公园内有野生大熊猫1340只，占全国野生大熊猫总数的71.89%；栖息地面积1.50万平方千米，占全国大熊猫栖息地面积的58.48%。该区域为亚热带季风气候区与青藏高寒气候区过渡带，地形复杂，生境多样，植被垂直带分布明显，孕育了丰富的生物多样性，是大熊猫及其伞护的8000多种野生动植物的最佳庇护所，极具核心保护价值。

大熊猫国家公园在现有已经建成的工程建筑和交通设施的基础上，通过建设空中廊道、地下隧道等方式，为大熊猫及其他动物留出通行通道，改善了大熊猫栖息地的碎片化现状，有助于实现隔离种群之间的基因交流。

Giant Panda National Park: 70% of Encircled Giant Panda Habitats

In 2017, China initiated the construction of the Giant Panda National Park. With a total area of 22,000 square kilometers, the Giant Panda National Park is one of the hotspots of global biodiversity conservation and the core distribution area of the wild population of giant pandas. The number of wild giant pandas was the largest in the Min Mountains, followed by the Qionglai Mountains, and the Daxiangling, Xiaoxiangling and Qinling Mountains were relatively small. There are 1,340 wild giant pandas in the Giant Panda National Park, accounting for 71.89% of the total wild giant pandas in China, with a habitat area of 15,000 square kilometers, accounting for 58.48% of the national giant panda habitat area. Located in the transition zone between the subtropical monsoon climate zone and the Qinghai-Xizang alpine climate zone, with complex terrain, diverse habitats, and obvious vertical vegetation distribution, the area has bred rich biodiversity, and is the best shelter for more than 8,000 species of wild animals and plants of giant pandas and its umbrella species, which has great core conservation value.

The Giant Panda National Park has set aside passage for giant pandas and other animals through the construction of aerial corridors and underground tunnels, which has reduced the fragmentation of giant panda habitats and helped to realize gene exchange between isolated populations.

秦岭片区
Qinling Mountains Area

白水江片区
Baishuijiang Area

岷山片区
Min Mountains Area

邛崃山-大相岭片区
Qionglai-Daxiangling Mountains Area

大熊猫国家公园四大片区
Four areas of the Giant Panda National Park

迁地保护：复壮大熊猫野生种群
Ex-situ Conservation: Wild Giant Panda Population Revitalization

迁地保护是相对于就地保护的另一种保护方式，是对就地保护方式的补充。迁地保护是指为了提高大熊猫的存活率，将大熊猫从野外转移到人工环境（如动物园、饲养场）或另一适宜生存的环境下进行保护。对濒危动物进行迁地保护的最终目的是为了重建或复壮野生自然种群，而不是为了用人工种群取代野生种群。

在迁地保护过程中，人工圈养大熊猫遭遇了很多方面的困难和挑战。经过几十年的探索，科研人员采用各种科技手段，对大熊猫的遗传、繁殖、疾病、营养等进行了深入研究，大大提升了圈养大熊猫种群的数量与质量。

Ex-situ conservation is an alternative to and complementary to in-situ conservation. *Ex-situ* conservation means to move and conserve giant pandas from the field to an artificial environment (like zoos and feedlots) or other habitable environment to improve the survival rate. *Ex-situ* conservation for endangered animals is ultimately to rebuild or restore wild natural populations, instead of replacing wild populations with artificial populations.

In the process of *ex-situ* conservation, captive breeding of giant pandas has encountered many challenges. Fortunately, the quantity and quality of captive giant panda population has been greatly improved after researchers conducted in-depth research on the inheritance, breeding, disease, and nutrition of giant pandas for decades by various technologies.

游客们在成都大熊猫繁育研究基地参观
Visitors at the Chengdu Research Base of Giant Panda Breeding

成都大熊猫繁育研究基地全景手绘图
Panoramic Freehand Drawing of the Chengdu Research Base of Giant Panda Breeding

大熊猫圈养小历程

按世界自然保护联盟（IUCN）推荐的标准，当野生动物种群数量不足1000只时，该动物种群就无法自然维持长期生存，这时就应该考虑建立圈养保护机构，进行人工饲养繁育。

建立大熊猫圈养保护机构，为危难中的大熊猫提供"庇护所"，能让它们在饮食、繁育、疾病等方面都得到良好的照顾。大熊猫圈养种群的壮大也为大熊猫将来回归野外家园打下了坚实的基础。

1953年1月17日，百花潭动物园（成都动物园的前身）救助了一只在都江堰市被发现的幼年大熊猫"大新"，这是新中国历史上第一只被救护的大熊猫，大熊猫迁地保护之路由此发源。

1963年，北京动物园的大熊猫通过自然交配产下一仔，这是世界上第一只在圈养环境中出生的大熊猫。1978年，北京动物园又首次成功实现了人工繁育大熊猫。这之后，成都、卧龙、重庆、福州、西安、上海、杭州和昆明等地的动物园和饲养研究机构也相继成功实现了人工繁育大熊猫。

History of Giant Pandas in Captivity

The International Union for the Conservation of Nature (IUCN) suggests that when the population of one wild animal is less than 1,000 and the population cannot sustain itself naturally in the long term, consideration should be given to the establishment of captive conservation institutions for captive breeding.

The captive giant panda conservation institutions are the sanctuaries for giant pandas under threat and care for them by providing food, helping them breed and fight off diseases. The expansion of the captive giant panda population lays a solid foundation for giant pandas to return to the wild.

On January 17, 1953, Baihuatan Zoo (now Chengdu Zoo) rescued Da Xin, a giant panda cub found in Dujiangyan. Da Xin was the first rescued giant panda in China, which marked the beginning of ex-situ conservation of giant pandas.

In 1963, a giant panda at the Beijing Zoo bred naturally and delivered a cub, the first giant panda born in the captive environment in the world. In 1978, Beijing Zoo succeeded in artificially breeding giant pandas for the first time. After that, zoos and breeding and research institutions in Chengdu, Wolong, Chongqing, Fuzhou, Xi'an, Shanghai, Hangzhou, and Kunming also succeeded in artificially breeding giant pandas.

"大新"的标本
Specimen of Da Xin

第一只在人工饲养条件下出生的大熊猫"明明"和它的母亲"莉莉"
Ming Ming, the first giant panda born in captivity, with his mother Li Li

大熊猫和花,小名花花,是成都大熊猫繁育研究基地最受欢迎的大熊猫之一。
He Hua, nicknamed Hua Hua, is one of the most popular giant pandas at the Chengdu Research Base of Giant Panda Breeding.

The Professional and Attentive Panda Conservationists

专业又用心的"熊猫人"

There is a group of panda conservationists who have been dedicating themselves to serving lovely giant pandas. Guess what? After 60 to 70 years of science-based conservation, the once endangered giant pandas are now witnessing an increase in the number of captive and wild populations, and "national parks" mark another start of their habitat protection. This means that great achievements have been made in this regard. But those fruitful results would not be possible without efforts of generations of panda conservationists for giant pandas. Their continuous exploration and dedication in field patrols, scientific research, and captive breeding have provided an amazing and admirable model for harmonizing the relationship between human beings and nature, and jointly building a shared future for all life on earth.

在可爱的大熊猫背后,有一群一直默默奉献的"熊猫人",你知道吗?经过六七十年的科学保护,一度濒危的大熊猫现在无论是圈养数量还是野外种群数量都出现了恢复性增长,栖息地保护也迈入了"国家公园"时代。这意味着大熊猫在物种保护方面取得了巨大成就!对大熊猫及其栖息地的保护,离不开一代代"熊猫人"的付出。他们在野外巡护、科学研究、人工饲养等方面的不断探索、全心投入,为努力协调人与自然的关系,共同构建地球生命共同体提供了一个让人惊叹和钦佩的范本。

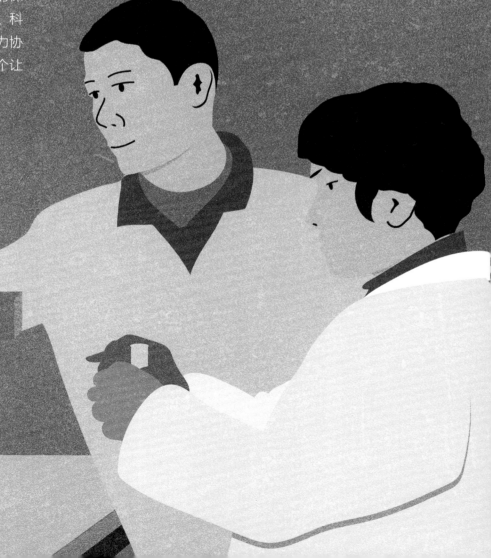

野外巡护：毒蛇猛兽常偶遇
Field Patrol: Poisonous Snakes and Beasts

The wild giant pandas live in areas with tall mountains and steep cliffs, deep valleys and dense forests, and frequent precipitation. Rangers patrol through remote mountainous areas at an altitude of several thousand meters, guarding the giant panda's home. They carry heavy devices and equipment on their back and trek through mountains and rivers to observe and record the changes in the species. They also need to retrieve and replace infrared cameras deep in the mountains.

In addition, rangers are also on the look out for venomous snakes and other beasts that may harm them. They are often punctured by bamboo and have their blood drunk by leeches, but they think it is worth it and take pleasure in what they do because their work is crucial to the conservation of the giant panda population.

野生大熊猫生活的地区山高崖陡、谷深林密、降水频繁。巡护人员在海拔几千米的无人山区巡回穿行，守护着大熊猫家园。他们背着沉重的仪器装备，跋山涉水、风餐露宿，每天在荒山密林中行走，沿途观察记录物种生存变化信息，还要取回安装在山上的红外相机，并更换新的红外相机。

此外，巡护队员们还要面对毒蛇、猛兽的袭击；而被竹子刺伤、蚂蟥噬咬则是"家常便饭"。但他们无怨无悔、以苦为乐，他们所做的工作意义非凡，为大熊猫种群保护奠定了坚实的基础。

寻找大熊猫粪便是巡护队员最关键的任务之一。通过分析大熊猫的粪便，可以帮助研究人员判断大熊猫的种群数量及栖息地范围。

Rangers' key task is to scour the wild for giant panda feces. Researchers then analyze the feces to determine the quantity and habitats of giant panda population.

科研工作：突破，再突破！
More Scientific Research Breakthroughs!

大熊猫科学研究与保护事业取得的阶段性成果，与一大批奋斗在一线的科研人员密不可分。在野外，他们追踪大熊猫足迹，趟过溪流，爬过山坡，穿过密林，研究大熊猫留下的痕迹、主食竹类、栖息环境……取得大熊猫研究的第一手资料，为生物多样性保护提出专业方案；在实验室里，他们反复做着实验，不断调整实验方案，在大熊猫保护应用基础研究方面取得了一项又一项突破。他们在大熊猫保护研究领域做出的巨大贡献，不仅仅保护了大熊猫种群，还保护了大熊猫栖息地的其他物种以及整个生态系统。

Scientists and researchers are laboring on the front line to contribute to the achievements of giant panda scientific research and conservation. They trek through creeks, hills, and forests; study the traces, staple bamboo and living environment of giant pandas to attain first-hand information and propose professional biodiversity conservation plans. In the laboratory, they have repeatedly done experiments and adjusted experimental protocols, making one breakthrough after another in the basic research on the application of giant panda conservation. They have made tremendous contributions to research on giant panda conservation, protecting not only the giant panda population, but also other species in the giant panda habitat and entire ecosystem.

设立于成都大熊猫繁育研究基地内的四川省濒危野生动物保护生物学重点实验室——省部共建国家重点实验室培育基地
Sichuan Key Laboratory of Conservation Biology of Endangered Wildlife established in Chengdu Research Base of Giant Panda Breeding — State Key Laboratory Cultivation Base Co-founded by Sichuan Province and the Ministry of Science and Technology of the People's Republic of China established in Chengdu Research Base of Giant Panda Breeding

饲养工作：奶爸、奶妈不轻松
Feeding Is Not an Easy Job

在圈养保护机构里，每只大熊猫都有专门的饲养员负责照料。大熊猫的饲养员既是"铲屎官"，又是大熊猫们的"奶爸""奶妈"。

大熊猫的"铲屎官"

大熊猫饲养员的日常工作非常繁杂。他们要对兽舍进行消毒和打扫，为大熊猫准备食物，清理和搜集大熊猫的粪便，观察大熊猫的精神、活动、采食、排便等情况并认真记录，对大熊猫进行行为训练，还要为大熊猫构建能满足它们生理和心理需求的生活环境，提升圈养大熊猫的福利与幸福感。

"奶爸"和"奶妈"

除了要操心大熊猫的日常起居，饲养员还要对大熊猫找对象、怀孕、产仔、育幼进行全程关照。尤其是在大熊猫分娩之际，他们总是彻夜不眠地守候在一旁。大熊猫们在饲养员的精心照料下慢慢长大，"结婚生子"。饲养员对于它们来说，就像"奶爸"和"奶妈"。

Every conservation agency for captive giant pandas assigns special keepers to take care of each giant panda. The keepers not only clean up after giant pandas, but also act as a surrogate parent to the giant pandas.

Feces cleaners

The keepers have many chores. They disinfect and clean the giant panda houses and prepare their food. They also collect giant panda feces and observe and record the spirit, activity, feeding, defecation, and other conditions of giant pandas. In addition, they conduct behavioral training for giant pandas, build a living environment that can meet the physical and psychological needs of giant pandas, and improve their welfare and happiness.

Surrogate parents

In addition to observing what the giant pandas do on a daily basis, the keepers also pay attention to their mating, pregnancy, delivery, and child rearing. When giant pandas give birth, the keepers stay up all hours near them. Giant pandas grow up under their watchful eyes and "get married and have children". Keepers act as their parents.

大熊猫饲养员的一天
A keeper's day

穿工作服，对兽舍区域进行消毒。

Wearing their uniform and disinfecting the giant panda houses.

观察大熊猫精神、活动、粪便、采食等情况，并认真填写动物饲养记录。

Observing their spirit, activity, feces, feeding, and other conditions, and recording feeding conditions.

为大熊猫添加食物。
Giving food to giant pandas.

打扫卫生，收集粪便。
Cleaning houses and collecting feces.

做好相应的行为观察和记录。
Observing and recording giant panda behaviors

做好动物行为训练及丰容工作，保证丰容材料及训练安全。
Conducting behavioral training, enriching the living environment, and ensuring that enrichment materials and training are safe.

Giant Panda Diseases and Treatment

大熊猫的疾病与救治

Sick? Yes, giant pandas can fall ill too. Some diseases may not only threaten their lives, but also bring about the demise of the entire giant panda family. In captivity, staff can ensure the health of giant pandas with the help of science-based breeding and the comprehensive health monitoring system. What are the common diseases affecting giant pandas? Will they obediently take medicine when they are sick? Let's find out.

生病？对，大熊猫也会生病。某些疾病一旦发生，不但可能威胁它们的生命，还可能给整个大熊猫家族带来灭顶之灾。在圈养条件下，工作人员可以通过科学饲养和完善的健康监测体系来保证大熊猫的身体健康。大熊猫会有哪些常见疾病呢？它们生病了会乖乖吃药吗？一起来了解一下吧。

Parasites

Scientists have found 22 species of parasites in giant pandas, and the most common parasitic diseases in giant pandas include the *ascariasis*, *psoroptic acariasis*, *demodicidosis*, and *ixodiasis*, etc.

Infectious diseases

Giant pandas are most susceptible to infectious diseases including bacteria diseases (such as diarrhea and hemorrhagic enteritis caused by *Escherichia coli*); viral diseases (such as canine distemper caused by canine distemper virus, and viral enteritis caused by the *parvovirus*); and fungal infectious diseases (such as hair loss caused by *Trichophyton*).

Internal medicine diseases

The most common internal medicine diseases threatening giant pandas are gastroenteritis, colds, pneumonia, and more.

Surgical diseases

The most common surgical diseases threatening giant pandas are trauma, fractures, head injuries, tumors, and more.

Obstetric diseases

The obstetric diseases that giant pandas face mainly include false pregnancy, miscarriage, dystocia, vulvar edema, endometritis, ovarian cysts, fallopian tube blockage, infertility, and more, among which false pregnancy and miscarriage are more common.

大熊猫易患这些病
>> Threatening Diseases

寄生虫病

科学家们已在大熊猫身上发现了22种寄生虫，大熊猫最常见的寄生虫病包括蛔虫病、痒螨病、蠕形螨病、蜱病等。

感染性疾病

大熊猫最容易患上的感染性疾病包括细菌性疾病，如大肠杆菌引起的腹泻、出血性肠炎等；病毒性疾病，如犬瘟热病毒引起的犬瘟热病，细小病毒感染导致的病毒性肠炎；真菌感染疾病，如毛癣菌引起的脱毛。

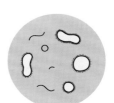

内科疾病

大熊猫最常见的内科疾病有慢性肠胃炎、感冒、肺炎等。

外科疾病

大熊猫最常见的外科疾病有外伤、骨折、颅脑损伤、肿瘤等。

产科疾病

大熊猫的产科疾病主要包括假孕、流产、难产、外阴水肿、子宫内膜炎、卵巢囊肿、输卵管堵塞、不育症等，其中假孕、流产较为常见。

Common diseases across different ages

Newborns: Digestive disorders, flatulence, ascites, etc.
Infant giant pandas: Pneumonia
Sub-adult giant pandas: Malnutrition
Adult giant pandas: Intestinal obstructions
Old giant pandas: Epilepsy, pneumonia, malnutrition syndromes

不同年龄段大熊猫的常见疾病

大熊猫幼仔： 消化功能紊乱、肠胀气、腹水等
幼年大熊猫： 肺炎
亚成体大熊猫： 营养不良
成年大熊猫： 肠梗阻
老年大熊猫： 癫痫、肺炎、营养不良综合征

大熊猫生病那些事儿
Stories about Sick Giant Pandas

"二丫头"生宝宝难产了

2004年，大熊猫"二丫头"就要生宝宝了。但是，它在第一次破羊水后14小时内都没表现出任何产前反应。专家们通过B超检查发现，胎儿还一直停留在"二丫头"的子宫内。最后，专家们通过保守治疗方法，将已死亡的胎儿取出母体，避免了外科手术，最大限度地保护了"二丫头"。这是全球范围内报道的首例大熊猫难产病例。

Er Yatou went through a difficult birth

In 2004, Er Yatou was on the verge of giving birth. However, it didn't show any prenatal reactions within the 14 hours after the water broke. experts did a B-scan and found that the fetus was lodged in Er Yatou's uterus and had perished. Finally, through conservative treatment, experts were able to remove the deceased fetus from the mother's body, avoiding surgery and protecting Er Yatou to the greatest extent. This is the first reported case of dystocia in a giant panda all over the world.

大熊猫也会得牙病

大熊猫以竹子为主食，这就决定了大熊猫必须拥有强健的牙齿。随着年龄的增加，大熊猫的牙齿会发生自然磨损，通过定期为大熊猫做牙齿检查，可以及时发现大熊猫的牙釉质磨损、牙髓暴露、牙结石、牙根尖周炎、牙本质染色等情况。大熊猫的牙齿护理和口腔医疗，已经成为大熊猫临床医学中的重要内容。

Tooth diseases

Bamboo is the giant panda's staple food, and this suggests they have rough teeth. As they age, the teeth are naturally worn down. Regular dental examinations of giant pandas have helped us timely detect enamel wear, pulp exposure, dental calculus, apical periodontitis, and dentin staining. Dental care and oral medicine for giant pandas have become an important part of the clinical treatments for giant pandas.

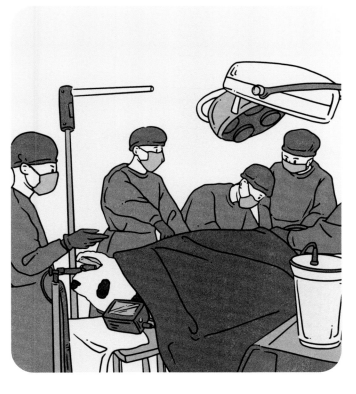

"胖美女"开刀取出了啥?

大熊猫"囡囡"身材丰满、胃口好。心宽体胖的它曾有过一段肠梗阻的不堪经历。2016年12月,"囡囡"出现烦躁不安、腹疼、频频举尾、排不出粪便等症状。经专家会诊,初步诊断为肠梗阻。通过手术,医生从"囡囡"的肠道中取出了堵塞的8个粪团,共重2.8千克。最终,"囡囡"在医护人员的精心护理下康复。"囡囡"开创了大熊猫"开刀取屎"的先例。

What has been removed during the surgery of a plump giant panda?

Nan Nan is a plump giant panda with a good appetite. However, this carefree panda had an unbearable experience of intestinal obstruction. In December 2016, Nan Nan had symptoms that included irritability, abdominal pain, frequent tail lifting, and inability to pass feces. After consulting experts, Nan Nan was diagnosed with an intestinal obstruction. The veterinarian was able to surgically remove 8 clogged fecal masses weighing a total of 2.8 kilograms from Nan Nan's intestines. Eventually, Nan Nan recovered under the care of the medical staff. Nan Nan set a precedent for giant pandas going under the knife for feces removal.

它感染了1605条蛔虫!

圈养条件下,大熊猫因为有良好的医疗护理,寄生虫病的感染率和感染强度往往很低,通常不会对生命安全构成严重威胁。而在野外条件下,大熊猫寄生虫病的感染率和感染强度都远高于圈养大熊猫,这已成为野生大熊猫消瘦和死亡的主要原因。成都大熊猫繁育研究基地抢救过的野外大熊猫蛔虫感染率几乎达到100%。在其中一只得到救治的野生大熊猫体内,研究人员发现了1605条蛔虫!

Infected with 1,605 roundworms!

Captive giant pandas have a low infection rate and intensity of parasitic diseases that do not pose a serious threat to their life because of prompt medical care. However, those in the wild have a much higher infection rate and intensity of parasitic diseases. This has become a major cause of emaciation and death in wild giant pandas. Almost 100% of the wild giant panda rescued by the Chengdu Research Base of Giant Panda Breeding were infected with roundworms. In one rescued wild giant panda, researchers found 1,605 roundworms!

大熊猫"缘小"被脱了"裤子"

大熊猫"缘小"因为一张后腿被剃掉毛发的照片爆红网络。2018年3月,饲养员发现大熊猫"缘小"右后肢无法着地,通过系列检查,兽医确诊其股骨骨折,决定对其进行手术治疗。为了保证手术的顺利进行,兽医剃掉了"缘小"的后腿毛发。在医护团队昼夜不停地精心护理下,"缘小"的精神和采食迅速恢复,伤口也很快愈合。

Yuan Xiao without pants

Yuan Xiao went viral because of a photo of its hind legs shaved off. In March 2018, the keeper discovered that Yuan Xiao was having issues with its right hind leg. After a series of examinations, Yuan Xiao was diagnosed with a femoral fracture and underwent surgery. To ensure the operation went smoothly, the veterinarian shaved the fur on Yuan Xiao's hind leg. Under around-the-clock care of the medical team, Yuan Xiao's spirit and appetite quickly recovered, and the wound healed quickly.

给大熊猫喂药是个技术活

成年大熊猫个子高、体重大,关键还力大无穷,强迫它们吃药几乎没可能,但是兽医们会先根据大熊猫的年龄、体重、病情等调节用药剂量,然后根据药的形态和口感,加入适量的糖,再偷偷把药混合在牛奶和水中,或夹在窝窝头、水果中,这样,大熊猫就会乖乖吃药啦!

Feeding medicine is a tricky task

It is nearly impossible to get a tall, heavy, and strong adult giant panda to take its medicine. Therefore, veterinarians will first adjust the dosage based on the age, weight, and condition of the giant panda, and then add an appropriate amount of sugar according to the form and taste of the medicine, and secretly mix the medicine in milk and water, or sandwich it in a panda cake (wowotou) or fruit. This is the only way to get a giant panda to take its medicine!

救治野生大熊猫："北川"的故事

Rescuing the Wild Giant Panda: Bei Chuan

If a sick giant panda is found in the wild, it can be treated on the spot by an experienced veterinarian in most cases. Individuals who are in poor physical condition and particularly weak will be sent somewhere with better medical facilities for treatment in a timely manner.

In 2010, some villagers found a seriously ill female giant panda in Honghe Village, Xiaoba Township, Beichuan County, Sichuan. Hearing that, the Chengdu Research Base of Giant Panda Breeding quickly dispatched an expert team to rescue the giant panda. After consultation, the experts found that the wild giant panda was in critical condition due to inflammation of its digestive system, which led to anemia and dehydration. To seize their chance to rescue the giant panda, the team decided to immediately transfer it to the Chengdu Panda Base for treatment, and the giant panda was thus given a new lease of life.

To commemorate this rescue experience, the giant panda was named Bei Chuan.

如果在野外发现了患病大熊猫，多数情况下可由经验丰富的兽医进行现场救治。其中体况不良、特别虚弱的个体，会被及时送至医疗条件较好的地方进行救治。

2010年，村民在四川北川县小坝乡洪河村发现了一只身患重病的雌性大熊猫。得到消息后，成都大熊猫繁育研究基地立即派出专家组前往进行抢救。经过会诊，专家们发现这只野生大熊猫因消化系统出现炎症导致了贫血、脱水症状，情况十分危急。为了抓住最佳抢救时间，专家组决定立即将这只野生大熊猫移送到成都熊猫基地接受治疗。这只大熊猫因此重获新生。

为了纪念这一段救护经历，它被取名为"北川"。

如今，重获新生的"北川"不仅胃口好，还出落得更加"标致"了。
Now, Bei Chuan not only has her good appetite, but also has become more beautiful.

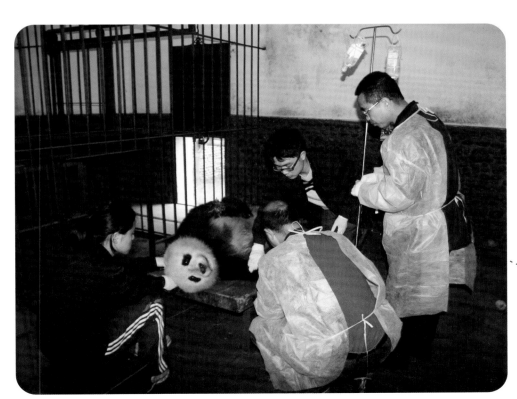

"北川"在接受救治
Bei Chuan receiving treatment

Giant Panda Rewilding

大熊猫野化放归

大熊猫的家在高山密林，那才是它们撒欢嬉戏、不负"熊生"之地。将圈养的大熊猫经野化训练后放归自然，是人工繁育的终极目标，也是降低局域小种群灭绝风险、复壮野外种群的重要手段。2003年，中国启动圈养大熊猫野化培训和放归研究。对圈养大熊猫进行野化培训后放归野外，可增加处于隔离生境单元中的大熊猫的数量，促进大熊猫之间的血缘交换，其种群的遗传多样性，这也是大熊猫迁地保护的核心目的。但是，由于大熊猫栖息地破碎、野外生存能力弱等原因，成功放归大熊猫可不是一件简单的事儿。

Giant pandas live in the mountains and dense forest, where they play and live up to their "bear life". Releasing captive giant pandas into the wild after rewilding is the ultimate goal of artificial breeding, and it is also an important means to reduce the extinction risk of small local populations and restore wild populations. In 2003, China launched captive giant panda rewilding training and reintroduction studies. Rewilding and releasing captive giant pandas can increase the number of giant pandas in isolated habitat units, promote blood exchange between giant pandas and the genetic diversity of their populations, which is also the core purpose of ex-situ conservation of giant pandas. However, due to the fragmentation of the giant panda's habitat and their weak survival ability in the wild, it is not easy to successfully release giant pandas.

从人工圈养到回归山林
From Captivity to the Wild

There are two main ways to rewild and release giant pandas in captivity "hard release" and "soft release". "Hard release" means to select suitable giant pandas and release them directly into nature after rewilding training, which requires no researchers to take care of them; "Soft release" means to undergo a series of artificially assisted training before they are released. After that, researchers provide continuous monitoring and assistance to the giant pandas, ensuring that they can timely know giant pandas' conditions in the wild and rescue them in case of danger. Let's take a look at the phases of human-assisted "soft releases".

Phase 1　Giant panda selection

Researchers select suitable giant pandas as trained individuals after a series of analyses and evaluations.

Phase 2　Rewilding training

Researchers will accompany giant pandas as they grow up. They build mutual trust and close ties with giant pandas through interaction. At the same time, giant pandas start to exercise in a simulated wild environment to acquire various survival skills required in the wild.

At the rewilding training stage, the giant panda will enter the rewilding training adaptation field to exercise its ability to forage for food, find a resting place, and avoid natural predators in the wild. After that, the giant panda will be trained in habitat adaptation and interactive training in the wild.

对圈养大熊猫进行野化放归主要有"硬放归"和"软放归"两种方式。"硬放归"是指选择合适的大熊猫个体，经过野化培训后直接放归自然，不需要研究人员进行照顾；"软放归"是指在放归前对大熊猫进行一系列人工辅助训练，放归后由研究人员给予大熊猫持续的监测和一定帮助，以保证能够及时了解其在野外的状况，若其遇到危险，则及时进行救治。让我们来看看人工辅助下的"软放归"包括哪几个阶段吧。

第一阶段　个体选择

研究人员经过一系列的分析和评估，挑选出合适的大熊猫作为培训个体。

第二阶段　野化培训

研究人员会陪伴大熊猫成长，通过互动来建立与大熊猫之间的相互信任和紧密联系。同时，让大熊猫在模拟野外的环境中开始锻炼自己，以获得野外环境下所需要的各项生存技能。

到了野化培训阶段，大熊猫将进入野化培训适应场，锻炼野外觅食能力、寻找卧息地的能力以及躲避天敌的能力等等。在此之后，便是对大熊猫进行栖息地适应与野外交互式培训。

检查野化培训个体的身体状况
Checking the physical condition of an individual in rewilding training

大熊猫在野化培训场接受训练
Giant panda being trained in the rewilding training field

第三阶段 放归前的准备

正式放归前，需要对大熊猫身体的各项指标进行评估。达标后，会邀请业内专家召开论证会，经评估确定野化大熊猫达到放归标准后再进入下一阶段。

与此同时，研究人员会通过分析野外调查数据，选择适宜的放归区域，之后再进行野外致危因子清除，并联合当地保护区组建专业的监测队伍，为大熊猫放归工作做好全面的准备。

第四阶段 自然放归与野外监测

最后，研究人员会将大熊猫带到提前选好的放归地点，进行自然放归。野化大熊猫回归大自然后，研究人员将通过它们佩戴的 GPS 项圈发出的信号，追寻它们的行踪，搜集它们的野外活动痕迹，比如粪便，用以进行 DNA 鉴定以便确定其身份；在进行野外调查监测的同时也会借助无人机等现代科技手段，对它们进行长期的野外监测。

第五阶段 大熊猫放归宣传教育

取得大熊猫放归地周边居民和社会公众的支持，是野化放归项目成功的重要前提。经过大家的不懈努力，大熊猫才能最终走向野外。

Phase 3 Preparation before release

The body index of giant pandas will be evaluated before release. After the index reaches the required standards, industry experts will be invited to evaluate and determine in a demonstration meeting that the rewilded giant pandas meet the release standards before entering the next phase.

Meanwhile, researchers will select appropriate release areas and eliminate field risk factors with the analysis of field survey data. Then, researchers establish a professional monitoring team with local reserves to make comprehensive preparations for the release of giant pandas.

Phase 4 Natural release and field monitoring

Finally, the researchers will take the giant pandas to a pre-selected release site for natural release. Once the giant pandas are released, researchers will track them through signals from the GPS collars they wear to find traces of their activities in the wild, such as feces, for DNA identification to determine their identity. During wild survey monitoring, researchers will also use modern scientific and technological means such as drones to conduct long-term field monitoring.

Phase 5 Publicity and education on the release of giant pandas

To win the support of the residents and the public near the giant panda release site is an important prerequisite for the success of the rewilding project. Through everyone's unremitting efforts, the giant panda can finally returning to the wild homeland.

科普教员在大熊猫放归地周边进行宣传教育
Conservation education instructor conducting publicity and education around the giant panda release site

佩戴了 GPS 项圈的大熊猫在雪地里活动
Giant pandas with GPS collars trekking through the snow

参加野化培训的大熊猫在野外觅食
Giant pandas under rewilding training foraging for food in the wild

大熊猫野外生存的必备技能
Essential Skills for Giant Pandas to Survive in the Wild

为了让大熊猫具备野外生存经验,科研人员会挑选动植物丰富的生态环境,在那里辅助它们克服各种困难,培养野外生存的必备技能。

自主觅食的能力

大熊猫以分布广泛、种类繁多的竹子为主食。在不同季节,它们会挑选不同种类的竹子,食用最新鲜、营养价值最高的部分,以满足自身的营养和能量需求。

寻找水源的能力

水源是野外生存的必需条件。大熊猫会在优质的水源附近活动。如何寻找深山里的水源也是它们必备的生存技能之一。

To equip giant pandas with experience in surviving in the wild, researchers will select ecological environments rich in animals and plants, where they can help giant pandas overcome various difficulties and develop the necessary skills to survive in the wild.

Foraging for food on its own

Giant pandas feed on bamboo which is widely distributed and varied. In different seasons, they select different species of bamboo and eat the freshest and most nutritious parts to meet their nutritional and energy needs.

Finding water

Water is essential for survival in the wild and giant pandas gravitate towards high-quality water sources. Finding water in the mountains is also one of their essential survival skills.

Marking territory

To mark resources and territories, giant pandas leave secretions and excreta in their natural environment. Male giant pandas urinate upside-down so that their urine can reach higher positions. Higher markings are more likely to attract mates because it will make females think that they are strong and tough.

Evading enemies

Danger lurks everywhere in the wild, and giant pandas must be able to protect themselves. Small juvenile giant pandas face more predators, so a high level of vigilance is essential. They have sensitive senses of hearing and smell, and once they sense danger, they will quickly hide or scramble up nearby trees.

标记领地的能力

为了标记资源和领地，大熊猫会在自然环境里留下分泌物和排泄物。雄性大熊猫以倒立的姿势小便，能够把尿液排到更高的位置。尿液标记的位置越高，意味着雄性大熊猫的体形越大，越能获得雌性大熊猫的青睐。

躲避敌害的能力

野外危机四伏，大熊猫必须具备保护自身安全的能力。体形较小的幼年大熊猫会面临更多的天敌，所以必须保持高度的警惕。它们的听觉和嗅觉非常灵敏，一旦感知到危险，就会迅速躲避，或者爬上附近的大树。

Downgraded from "endangered" to "vulnerable"

从濒危到易危

受威胁等级
Threat level

濒危（EN）
Endangered

1114只

20 世纪 80 年代
大熊猫野外种群数量
Wild Giant Panda Population Size in the 1980s

受威胁等级
Threat level

易危（VU）
Vulnerable

近 Nearly
1900只

今天
大熊猫野外种群数量
Current wild population of giant pandas

Exciting Conservation Effects

On September 4, 2016, the International Union for Conservation of Nature (IUCN) announced that the giant panda was downgraded from endangered to vulnerable. Up until then, the IUCN Red List of Threatened Species had classified giant pandas as endangered. According to the data from the fourth survey on wild giant pandas, the IUCN expert team assessed that the wild giant panda population was growing, and their habitat was recovering and decided to downgrade their level of endangerment.

Being downgraded from endangered to vulnerable reflects the remarkable results achieved in China's work in biodiversity conservation and ecological environment restoration. However, that does not mean giant pandas are no longer protected or valued.

Currently, giant pandas are still facing survival crises in all aspects. Although the quantity of wild giant pandas in China has reached nearly 1,900, they are fragmented into 33 local populations, 24 of which are at high risk of extinction. Thus, we should not only focus on the growth of giant panda populations, but also explore the establishment of a conservation system for endangered species, led by giant pandas. On this basis, we will promote the protection of the entire ecosystem and the harmonious development of mankind and nature.

Vulnerable species are also threatened. From vulnerable to near threatened or even least-concerned, there is still a long way to go to conserve giant pandas.

2016年9月4日，世界自然保护联盟（IUCN）宣布将大熊猫的受威胁程度从濒危降为易危。在此之前，IUCN濒危物种红色名录中一直将大熊猫划分在濒危等级。但是，根据中国第四次大熊猫野生资源调查的数据，IUCN专家组评估认为大熊猫的野生种群数量在增长，栖息地在恢复，所以决定将它们的濒危等级降级。

大熊猫从濒危变成易危，体现了我国在生物多样性保护与生态环境修复等方面所做的工作取得了明显的成效。但是，这并不意味着大熊猫将不再受到保护和重视。

当前，大熊猫的生存仍然面临着各方面的危机。虽然全国野生大熊猫数量已近1900只，但这些大熊猫被割裂成33个局域种群，有24个局域种群具有较高的灭绝风险。因此，我们不能只着眼于关注大熊猫数量上的增长，而应该探索建立以大熊猫为首的濒危物种保护体系，在此基础上，推动整个生态系统的保护和人与自然的和谐发展。

易危物种同样属于受威胁物种。从易危到近危甚至无危，保护大熊猫还有很长的路要走。

物种濒危等级

世界自然保护联盟濒危物种红色名录将物种濒危等级划分为9个，由高到低分别为：已灭绝、野外灭绝、极度濒危、濒危、易危、近危、无危、数据缺乏、未予评估。这其中，极危、濒危和易危物种又被统称为受威胁物种。

Category of endangered species

The IUCN Red List of Threatened Species classifies endangered species into 9 levels of endangerment: extinct, extinct in the wild, critically endangered, endangered, vulnerable, near threatened, least concern, data deficient and not evaluated from highest to lowest. Critically endangered, endangered, and vulnerable species are collectively referred to as threatened species.

受威胁 Be Under Threat

| 未予评估（NE）Not Evaluated | 数据缺乏（DD）Data Deficient | 无危（LC）Least Concern | 近危（NT）Near Threatened | 易危（VU）Vulnerable | 濒危（EN）Endangered | 极危（CR）Critically Endangered | 野外灭绝（EW）Extinct in the Wild | 灭绝（EX）Extinct |

灭绝风险 Risk of Extinction

Chapter 6　第六章
Taking Actions
生态家园

物我平等，共睦家园。
生态文明，深入人心。

Promoting equality between animals and humans and building a harmonious home. Ecological civilization is widely embraced by people.

我们大熊猫家族的野外栖息地风景可美了！那里有高山密林，还有清澈的流水，云雾缭绕，仿佛仙境一般。我们喜欢独居，但不等于我们不需要朋友。在我们的生态家园里，还生活着川金丝猴、小熊猫、羚牛、黑熊、红腹锦鸡、黄喉貂、云豹等动物邻居，生长着珙桐、巴山榧树、红豆杉、篦子三尖杉、西康玉兰、桢楠等植物朋友。我们彼此之间遵从着既定的生存法则，各自占有自己的生存空间，快乐和谐地生活在一起。走过漫长岁月，我们发现了一个有意思的现象：生态家园的成员越丰富、越多样，地球就越生机盎然！

>>　　We inhabit areas with breath-taking scenery of high mountains, dense forests, and crystal-clear flowing water, which are also shrouded in mist and cloud, like a fairy tale. We like to live alone, but that doesn't mean we don't need friends. We share our ecological homeland with some animals such as golden snub-nosed monkeys, red pandas, takins, black bears, golden pheasants, yellow-throated martens, and clouded leopards, as well as plants like *Davidia involucrata*, *Torreya fargesii*, *Taxus wallichiana*, *Cephalotaxus oliveri*, *Magnolia wilsonii*, and *Phoebe zhennan*. We obey the established laws of survival among each other, never intruding into each other's living space, and living happily and harmoniously together. After a long time, it strikes us that a more diversified ecology means a more vibrant earth!

嗨，我是小川。
Hi, I'm Xiao Chuan.

\>\>

Chengdu
Giant Panda
Museum
Exhibition Hall 6
Taking Actions

成都大熊猫博物馆 第六展厅 生态家园

Biodiversity in China
中国的生物多样性

生物多样性是地球上所有生命多样性的总和，包括遗传多样性、物种多样性和生态系统多样性。如果生物多样性被破坏，就像多米诺骨牌倒下了一块，会产生连锁效应，包括人类在内的其他"牌"也可能会先后倾倒。

Biodiversity is the sum of the diversity of all life on Earth, including genetic diversity, species diversity, and ecosystem diversity. If biodiversity is destroyed, like a domino falling by one, there will be a knock-on effect, and other "cards", including mankind, may also be toppled one after the other.

多样的生物，多样的精彩
>> Wonderfully Diverse Creatures

China has vast lands and ocean, diverse landforms and a varied climate, giving birth to a wide variety of ecosystems such as forests, grasslands, deserts, wetlands, and oceans.

China is home to about 150,000 species of insects, 35,000 species of other invertebrates, and more than 6,500 species of vertebrates, creating a rich and colorful animal world. Among them, more than 470 terrestrial vertebrates are endemic to China, such as giant pandas, crested ibises, snub-nosed monkeys, South China tigers, and Chinese alligators.

China also houses nearly one-tenth of the 300,000 species of plants on Earth. Among them, more than 15,000 species are endemic to China, such as *Ginkgo biloba*, *Metasequoia glyptostroboides*, *Pseudolarix amabilis*, *Davidia involucrata*, *Camptotheca acuminata*, and *Michelia odora*. These plants constitute a marvelous plant paradise.

However, China's biodiversity is increasingly threatened by habitat loss and fragmentation, over exploitation of resources, environmental pollution, invasive species, and global climate change. China is one of the first countries to accede to the *Convention on Biological Diversity*. In addition to that, to maintain biodiversity and ecological balance, China has also released the *National Outline for Ecological Environmental Protection* and *China's Biodiversity Conservation Strategy and Action Plan* and established the *Law of the People's Republic of China on the Protection of Wildlife* to expand the construction of nature reserves. After more than 60 years of efforts, China has established a system of natural protected areas at all levels and types, including pilot areas of the national park system, nature reserves, scenic spots, geological parks, forest parks, wetland parks, desert parks, marine special protected areas, with a total of 11,800 protected areas, accounting for 18% of the land area and 4.6% of the territorial sea. These protected areas play an important role in conserving biodiversity - 85% of China's wild animal populations and 65% of higher plant communities.

中国拥有辽阔的陆地和宽广的海洋，地貌多样，气候多变，孕育了森林、草原、荒漠、湿地和海洋等类型繁多的生态系统。

中国约有昆虫15万种，其他无脊椎动物3.5万种，脊椎动物6500多种，构建了丰富多彩的动物世界。其中，大熊猫、朱鹮、金丝猴、华南虎、扬子鳄等470多种陆栖脊椎动物更是仅分布于中国的特有物种。

地球上的植物大约有30万种，近十分之一生长在中国，其中15000多种植物是中国特有的，如银杏、水杉、金钱松、珙桐、喜树、观光木等，它们共同构成了一个神奇的植物天堂。

然而，由于生境丧失和破碎化、资源过度开发利用、环境污染、入侵物种及全球气候变化等因素，中国的生物多样性受到了日益严重的威胁。中国是最早加入《生物多样性公约》的国家之一。为了维护生物多样性和生态平衡，中国发布了《全国生态环境保护纲要》《中国生物多样性保护战略与行动计划》，制定了《中华人民共和国野生动物保护法》，并进一步加大了对自然保护地的建设。经过60多年的努力，初步建立起了包括国家公园体制试点区、自然保护区、风景名胜区、地质公园、森林公园、湿地公园、沙漠公园、海洋特别保护区等在内的各级各类自然保护地体系，各类保护地总数达1.18万个，占陆域国土面积的18%、领海面积的4.6%，有效保护了中国85%的野生动物种群、65%的高等植物群落，在保护生物多样性方面发挥了重要作用。

15000+
中国特有植物
Plants endemic to China

11800+
各级各类自然保护地
Nature reserves at all levels and types

470+
仅分布于中国的陆栖脊椎动物
Areas for terrestrial vertebrates only in China

The Importance of Conserving Giant Pandas
保护大熊猫的重要意义

Monolithic efforts are being made for giant panda conservation because of their cuteness and their unique position in the ecosystem. The southwest mountains of China at an altitude span of 500~7000 meters, from the plains to the snow-capped mountains, where giant pandas live, has nearly one-third of China's higher plants and nearly half of China's birds and beasts. It is a "biological gene bank" with the world's eyes on it. It is also one of 36 global biodiversity hotspots. As a species much older than mankind, the giant panda has long been inextricably linked to the surrounding ecological environment in its natural evolution. The species has become the flagship species and umbrella species of wildlife conservation in China, which is of great significance for the conservation of wild animals and plants.

花巨大的精力保护大熊猫并不仅仅是因为它们的呆萌可爱，更重要的原因是它们在生态体系中的独特地位。在大熊猫生活的中国西南山地，从平原到雪山500~7000米的海拔跨度，生长着中国近三分之一的高等植物，生活着中国近二分之一的鸟兽。这里是世界瞩目的"生物基因库"，36个全球生物多样性热点区域之一。作为比人类古老得多的物种，大熊猫在自然进化中早已和周边的生态环境构成了千丝万缕的共生关系，成为我国野生动物保护的旗舰物种和伞护种，对于野生动植物保护有着非常重要的意义。

旗舰物种：生态保护代言人
Flagship Species: Representative of Ecological Conservation

The flagship species refers to a species that has special appeal and attraction to ecological conservation, promoting societal attention to species conservation. It is a representative species of regional ecological maintenance. The survival of such a species generally does not have a serious impact on maintaining the integrity and continuity of ecological processes or food chains, but it has won the people's love and attention because of its charming appearance or other features (e.g. giant pandas, Siberian tigers, snow leopards). Therefore, the conservation of these animals can easily raise more funds that can be used to protect large-scale ecosystems. In 1961, when the World Wide Fund for Nature (WWF) was founded, the founder Sir Peter Scott designed the emblem based on the only giant panda in the Western world at the time, Ji Ji, officially establishing the leadership of the giant panda among the flagship species.

旗舰物种指某个物种对生态保护具有特殊号召力和吸引力，可促进社会对物种保护的关注，是地区生态维护的代表物种。这类物种的存亡一般对保持生态过程或食物链的完整性、连续性无严重的影响，但其魅力（外貌或其他特征）赢得了人们的喜爱和关注（如大熊猫、东北虎、雪豹等）。对这类动物的保护容易募集到更多的资金，从而用于保护大规模的生态系统。1961年，世界自然基金会（WWF）创办之际，创始人斯科特爵士就是以当时西方世界唯一的一只大熊猫"姬姬"为原型设计的会徽，正式确立了旗舰物种中大熊猫的领袖地位。

大熊猫
Giant panda

东北虎
Siberian tiger

雪豹
Snow leopard

伞护种：物种保护的最优选
Umbrella Species: Best Choice of Species Conservation

伞护种指那些自身的生存环境需求也能满足很多其他物种生存环境需求的物种。比如，鲸的生存环境需求涵盖了许多水生动物的生存环境需求，鲸就是一个伞护种。通过对该物种的保护，同时也能为很多其他物种提供保护。"伞护种"的概念最早是由美国人布鲁斯·威尔科克斯于1984年提出的。通过对伞护种的保护来保护更多物种，是人类在物种保护资金有限的情况下做出的最优化选择。

大熊猫是伞护种中最典型的物种，保护了大熊猫，就能保护大熊猫栖息地内众多的其他物种。

Umbrella species refer to those whose own habitat needs can also meet the habitat needs of many other species. For example, the living environment of whales can meet the needs of many aquatic animals. Therefore, the whale is an umbrella species. Protecting this specie can also provide protection for many other species. The concept of "umbrella species" was first proposed by Bruce Wilcox, an American, in 1984. Protecting more species through the protection of umbrella species is the optimal choice made by humans in the context of limited funding for species conservation.

The giant panda is the most typical umbrella species, so protecting this species can protect many other species in its habitats.

林麝
Forest musk deer

美容杜鹃
Rhododendron calophytum

斑羚
Goral

藏酋猴
Tibetan macaque

珙桐
Davidia involucrata

川金丝猴
Golden snub-nosed monkey

雪豹
Snow leopard

小熊猫
Red panda

黄喉貂
Yellow-throated marten

水青树
Tetracentron sinense

绿尾虹雉
Chinese monal

红腹锦鸡
Golden pheasant

四川花楸
Sorbus setschwanensis

毛冠鹿
Tufted deer

羚牛
Takin

红豆杉
Taxus chinensis

Giant Panda's Flora and Fauna Neighbors
大熊猫的动植物邻居

From giant panda nature reserves to the Giant Panda National Park, in addition to the giant panda, the umbrella species, there are more than 8,000 species of wild animals and plants such as Snow leopards, Golden snub-nosed monkeys, Chinese monals, Crested ibises, *Davidia involucrata*, and *Taxus chinensis*. All living things are coexisting in harmony. In today's giant panda ecological home, a beautiful picture of harmonious coexistence between man and nature has slowly unfolded.

从大熊猫自然保护区,到大熊猫国家公园,在这个以大熊猫为伞护种的生态家园里,除了大熊猫,还生活着雪豹、川金丝猴、绿尾虹雉、朱鹮、珙桐、红豆杉等 8000 多种野生动植物。万物生灵,美美与共,在如今的大熊猫生态家园里,一幅人与自然和谐共生的美好画卷已徐徐展开。

Sympatric Companion Animals of Giant Pandas

大熊猫的同域伴生动物

Red panda

English name: Red panda
Scientific name: *Ailurus fulgens*
Classification: Carnivora - Ailuridae - *Ailurus*
Lifespan: About 12.5 years old
Food: Red pandas love bamboo shoots, tender twigs, and bamboo leaves. They also eat wild fruits, leaves, moss, bird eggs, and sometimes small birds and other small animals and insects. They particularly like sweet food.
Appearance: The reddish-brown animals are about 80 ~ 120 centimeters in length, with white markings on their cheeks. They have a long and fluffy tail with 9 ~ 12 red and white ring patterns.
Distribution: They inhabit alpine jungles at an altitude of 2,500 ~ 4,800 meters in China, India, Laos, Nepal, the Kingdom of Bhutan, and Myanmar in the Himalaya-Hengduan mountain range. China has nearly 3,000 wild red pandas in Sichuan, Yunnan, and Xizhang, and there are fewer than 10,000 of them worldwide.
Protection class: Endangered species on the IUCN Red List, and the wildlife under state second-class key protection in China.

小熊猫

英文名： Red panda
拉丁学名： *Ailurus fulgens*
分类： 食肉目 – 小熊猫科 – 小熊猫属
寿命： 约 12.5 岁
食物： 喜食竹笋、嫩枝和竹叶，也吃野果、树叶、苔藓、鸟蛋，有时会捕食小鸟和其他小动物、昆虫，尤其喜食带有甜味的食物。
外形特点： 体长约 80 ~ 120 厘米，全身红褐色，脸颊有白色斑纹，尾巴长而蓬松，上有 9 ~ 12 个红白相间的环纹。
分布： 栖居于海拔 2500 ~ 4800 米的高山丛林地带，目前仅分布在喜马拉雅 – 横断山脉区域的中国、印度、老挝、尼泊尔、不丹和缅甸等国。在中国，小熊猫分布在四川、云南和西藏，野生数量约为 3000 只。它们在全球的数量已经不足 10000 只。
保护等级： IUCN 红色名录濒危物种，中国国家二级重点保护野生动物。

我大部分时间生活在树上

小熊猫性格温顺，机警胆小。它们早晚出来活动觅食，白天多在洞里或大树的阴凉处睡觉。小熊猫善于攀爬，在野外，它们大部分时间生活在树上。野生小熊猫成年后，在繁殖期和育幼期之外都过着独居生活。

I spend most of my time in trees.

The red panda is meek, alert, and timid. They scurry out in the morning and evening to forage for food, and during the day, they mostly sleep in their burrows or in the shade of large trees. They are adept climbers. In the wild, they spend most of their time in trees. When wild red pandas reach adulthood, they live a solitary life outside of the breeding and rearing periods.

小熊猫的分布范围
Distributed areas of red pandas

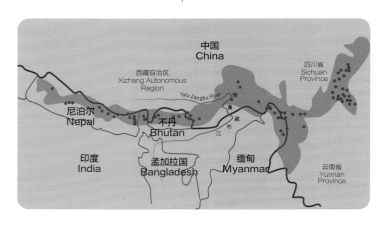

I can never be a giant panda even if I grow up.

Although red pandas and giant pandas have similar names and live in the same habitat and feed on bamboo, they are completely different from each other.

长大后我也成不了大熊猫

小熊猫和大熊猫虽然名字相近，并且在野外生活在同一片栖息地中，还都以竹子为主食，但它们是两种完全不同的动物。

	小熊猫	大熊猫
分类	小熊猫科	熊科
初生时体色	灰白色	粉色带稀疏白毛
成年后毛色	棕褐色	黑白相间
成年后体重	5千克左右	80~120千克
平均寿命	12.5岁	20~25岁
分布范围	中国、尼泊尔、缅甸、越南	中国
濒危等级	濒危	易危

	Red panda	Giant panda
Classification	Ailuridae	Ursidae
Color when born	White-gray	Pink with sparse white furs
Color in adulthood	Reddish-brown	Black and white
Weight in adulthood	About 5 kilograms	80~120 kilograms
Average lifespan	12.5 years old	20~25 years old
Distributed area	China, Nepal, Myanmar, Vietnam	China
Endangered category	Endangered	Vulnerable

Golden snub-nosed monkey

English name: Golden snub-nosed monkey
Scientific name: *Rhinopithecus roxellana*
Classification: Primates - Cercopithecidae - *Rhinopithecus*
Lifespan: 20 ~ 25 years old
Food: They like lichen, tender leaves and twigs, bark, flowers, fruits, etc.
Appearance: Their nostrils are tilted upward, which is commonly known as a "sky-facing nose". They have a slightly blue face and large areas of golden-brown fur.
Distribution: They inhabit forests at a high altitude of 1,500 ~ 3,300 meters in Sichuan, Shaanxi, Gansu, and Hubei. There are about 25,000 golden snub-nosed monkeys at present.
Protection class: Endangered species on the IUCN Red List, and the wildlife under state first-class key protection in China.

川金丝猴

英文名： Golden snub-nosed monkey
拉丁学名： *Rhinopithecus roxellanae*
分类： 灵长目 - 猴科 - 仰鼻猴属
寿命： 20 ~ 25 岁
食物： 喜食地衣、嫩叶、嫩枝、树皮、花、果等。
外形特点： 鼻孔向上仰，俗称"朝天鼻"。颜面部为淡蓝色，身上长有大面积的金棕色毛发。
分布： 栖息于海拔 1500 ~ 3300 米的高海拔地区的森林中，主要分布于四川、陕西、甘肃、湖北等地，目前种群数量约为 25000 只。
保护等级： IUCN 红色名录濒危物种，中国国家一级重点保护野生动物。

We do not have the monkey king.

The golden snub-nosed monkey is a social animal, and each large group is based on a small family with a wealth of social behaviors. There are two household types: one is composed of one male golden snub-nosed monkey and 3 ~ 5 female and baby monkeys; the other is composed of all male monkeys, commonly known as a "bachelor" family. There is no monkey king but a patriarch. In the first household type, the only male monkey is bigger and stronger than female ones and holds dominant position. Adult females also have a social hierarchy, while juvenile individuals are at the lowest rung.

这个猴群没有猴王

川金丝猴是群居动物，每个大的集群以家族性的小集群为活动单位，有丰富的社群行为。川金丝猴的家庭组合形式一般分为两种：一种由 1 只雄猴、3~5 只雌猴及幼猴组成；另一种全部由雄猴组成，俗称"单身汉"家庭。川金丝猴没有猴王，只有家长，在一雄多雌的川金丝猴族群中，只有一只雄性个体占优势地位。雄性比雌性个体要大得多，强壮得多。成年雌性个体也会有社会等级地位，幼年个体的地位最低。

川金丝猴分布范围
Distributed areas of golden snub-nosed monkeys

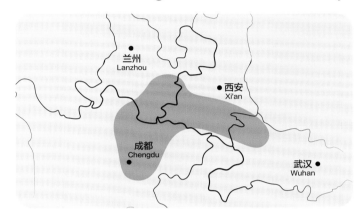

"金丝猴"是仰鼻猴属动物的统称。仰鼻猴属有 5 种动物，其中，只有川金丝猴的毛发是金黄色的。

"Snub-nosed monkey" is a generic term for animals of the genus *Rhinopithecus*. There are 5 species in the genus, of which only the golden snub-nosed monkey has golden-colored hair.

金丝猴大家族
Snub-nosed monkeys

滇金丝猴 Black snub-nosed monkey

黔金丝猴 Grey snub-nosed monkey

川金丝猴 Golden snub-nosed monkey

越南金丝猴 Tonkin snub-nosed monkey

缅甸金丝猴 Myanmar snub-nosed monkey

雪豹

英文名: Snow leopard
拉丁学名: *Panthera uncia*
分类: 食肉目 – 猫科 – 豹属
寿命: 8~13 岁
食物: 以岩羊、北山羊、盘羊等大型高原动物为主食,也捕食旱獭、鼠兔、野兔,以及雪鸡、马鸡、虹雉等小型动物,在食物缺乏时也盗食家畜、家禽。
外形特点: 全身灰白色,布满黑斑。头部黑斑小而密,背部、体侧及四肢外缘形成不规则的黑环,越往体后黑环越大,尾长而粗大。
分布: 从中亚至青藏高原和蒙古高原面积广袤的山地,包括中亚 12 国,印度也有分布。中国是雪豹数量最多、栖息地面积最大的国家。
保护等级: IUCN 红色名录易危物种,中国国家一级重点保护野生动物。

灰白色黑斑皮毛能帮我在雪地里很好地隐蔽自己。

Snow leopard

English name: Snow leopard
Scientific name: *Panthera uncia*
Classification: Carnivora - Felidae - *Panthera*
Lifespan: 8 ~ 13 years old
Food: They feed on large plateau animals such as rock goats, ibex, and argali. They also capture marmots, pikas, hares, and small animals such as snow cocks, eared pheasants, and rainbow pheasants. They also steal livestock and poultry when food is scarce.
Appearance: The grayish-white animal is covered with dark spots. The spots on the head are small and dense, and irregular black rings form on the back, side of the body, and the outer edges of the limbs. The black rings become larger on the posterior part of the body, and the tail is long and thick.
Distribution: The snow leopard is distributed in the vast mountainous areas from Central Asia to the Qinghai-Tibet Plateau and the Mongolian Plateau, including 12 countries in Central Asia, and India. China is home to the most snow leopards as well as their widest habitats.
Protection class: Vulnerable species on the IUCN Red List, and the wildlife under state first-class key protection in China.

Our grayish-white fur with dark spots can help hide ourselves in the snow.

Leopard

English name: Leopard
Scientific name: *Panthera pardus*
Classification: Carnivora - Felidae - *Panthera*
Lifespan: 14 ~ 19 years old
Food: It mainly preys on ungulates, but also on monkeys, rabbits, murids, birds, and fish. In autumn, it also eats sweet berries.
Appearance: They have yellow fur and black ring spots. The spots on the head is small and dense and large and dense on the back. The spots are round or oval in the shape of plum blossoms.
Distribution: It is the most widespread felid worldwide in Eurasia and Africa. It lives in forests, bush wood, wetlands, and desert. Leopards are scattered in the northeast, north, and southwest of China, and the southern slopes of the central Himalayas.
Protection class: Vulnerable species on the IUCN Red List, and the wildlife under state first-class key protection in China.

豹

英文名：Leopard
拉丁学名：*Panthera pardus*
分类：食肉目 – 猫科 – 豹属
寿命：14 ~ 19 岁
食物：主要捕食有蹄类动物，也捕食猴、兔、鼠类、鸟类和鱼类，秋季还采食甜味浆果。
外形特点：被毛黄色，满布黑色环斑，头部斑点小而密，背部斑点密而大，斑点为圆形或椭圆形的梅花状图案。
分布：世界上分布范围最广的猫科动物，分布区域横跨亚欧大陆与非洲大陆，生活在森林、灌丛、湿地、荒漠等环境中。在中国，散布在东北、华北、西南及喜马拉雅山脉中段南坡。
保护等级：IUCN 红色名录易危物种，中国国家一级重点保护野生动物。

> 我不仅吃地上跑的、水里游的，连甜味浆果我也喜欢。
>
> Not only do I eat terrestrial and aquatic animals, but I also like sweet berries.

金猫

英文名： Asian golden cat
拉丁学名： *Catopuma temminckii*
分类： 食肉目 – 猫科 – 金猫属
寿命： 10～20 岁
食物： 林麝、小麂、毛冠鹿等中小型偶蹄类植食动物，啮齿类、鸟类和爬行类动物。
外形特点： 体毛多为棕红或金褐色，两眼内侧各有一条白纹，额部有带黑边的灰色纵纹，延伸至头后。
分布： 主要分布于东南亚、中南半岛和中国的热带和亚热带湿润常绿阔叶林、混合常绿山地林和干燥落叶林，在中国，近年来仅在四川、陕西、云南和西藏有记录。
保护等级： IUCN 红色名录近危物种，中国国家一级重点保护野生动物。

Asian golden cat

English name: Asian golden cat
Scientific name: *Catopuma temminckii*
Classification: Carnivora - Felidae - *Catopuma*
Lifespan: 10 ~ 20 years old
Food: They eat small and medium-sized artiodactyl herbivores such as forest musk deer, muntjac, tufted deer, rodents, birds, and reptiles.
Appearance: They have brownish-red or golden-brown fur with a white stripe on the inside of each eye, and a gray longitudinal stripe with a black edge on the forehead that extends to the back of the head.
Distribution: It is mainly distributed in tropical and subtropical humid evergreen broad-leaved forests, mixed evergreen montane forests, and dry deciduous forests in southeast Asia, the Indochina Peninsula, and China. It has been seen in Sichuan, Shaanxi, Yunnan, and Xizang of China in recent years.
Protection class: Near Threatened species on the IUCN Red List, and the wildlife under state first-class key protection in China.

> 我头上的条纹超帅气！
> The stripes on my head are so cool!

Yellow-throated marten

English name: Yellow-throated marten
Scientific name: *Martes flavigula*
Classification: Carnivora - Mustelidae - *Martes*
Lifespan: Around 14 years old
Food: It eats birds, bird eggs, frogs, reptiles, insects, and plant fruits. A group of yellow-throated martens can hunt beasts bigger than them, like musk deer and deer.
Appearance: It was named because of the distinct yellow-orange laryngeal spots on the anterior chest. It has a slender and slightly triangular head, short and rounded ears, and a slender figure.
Distribution: It is mainly distributed in the forested areas of East and Southeast Asia and eastern Russia. In China, it is mainly distributed in the vast areas of Northeast China, Central China, South China, East China, and Southwest China, as well as the 2 large islands of Taiwan and Hainan.
Protection class: Least Concern species on the IUCN Red List, and the wildlife under state second-class key protection in China.

黄喉貂

英文名：Yellow-throated marten
拉丁学名：*Martes flavigula*
分类：食肉目 – 鼬科 – 貂属
寿命：14 岁左右
食物：鸟类、鸟蛋、蛙类、爬行类、昆虫和植物果实等，还能合群捕杀麝、鹿等比自身体形大很多的兽类。
外形特点：因前胸部具有明显的黄橙色喉斑而得名，头较为尖细，略呈三角形，耳短而圆，体形细长。
分布：主要分布于东亚和东南亚及俄罗斯东部林区，在中国，主要分布于东北、华中、华南、华东、西南的广大地区，以及台湾、海南 2 个大型岛屿。
保护等级：IUCN 红色名录无危物种，中国国家二级重点保护野生动物。

我们合群能捕杀比我们自己体形大很多的麝、鹿等兽类。
We together can hunt beasts bigger than us, like musk deer and deer.

羚牛

英文名： Takin
拉丁学名： *Budorcas taxicolor*
分类： 偶蹄科 – 牛科 – 羚牛属
寿命： 12～15岁
食物： 竹笋、竹叶、各种植物的枝叶与树皮等。
外形特点： 全球有4种羚牛，与大熊猫生活在同一片栖息地的有2种：一种是秦岭羚牛，全身毛色金黄；另一种是四川羚牛，毛色为棕黄色并夹杂大量的黑色斑块。它们都是中国特有的羚牛。
分布： 秦岭羚牛仅分布在陕西南部的秦岭山脉；四川羚牛分布于青藏高原东缘的山地，包括甘肃南部和四川北部至中部。
保护等级： IUCN红色名录易危物种，中国国家一级重点保护野生动物。

Takin

English name: Takin
Scientific name: *Budorcas taxicolor*
Classification: Even-toed ungulate - Bovidae - *Budorcas*
Lifespan: 12 ~ 15 years old
Food: It eats bamboo shoots and leaves as well as the branches and bark of various plants.
Appearance: There are 4 species of takins, two of which live in the same habitat with giant pandas: one is the golden takin with golden fur and the other is Sichuan takin with brownish-yellow fur and lots of black patches. Both are endemic to China.
Distribution: Golden takins are distributed in the Qinling Mountains in southern Shaanxi; Sichuan takins are in the mountains of the eastern edge of the Qinghai-Tibet Plateau, including southern Gansu and northern to central Sichuan.
Protection class: Vulnerable species on the IUCN Red List, and the wildlife under state first-class key protection in China.

想不到吧，我也吃竹子。
Guess what? I eat bamboo, too.

Chinese monal pheasant

English name: Chinese monal
Scientific name: *Lophophorus lhuysii*
Classification: Galliformes - Phasianidae - *Lophophorus*
Lifespan: Unknown
Food: They feed mainly on the shoots, leaves, fruits, seeds, and rhizomes of plants.
Appearance: Males have purple crests with a white lower back and blue-green tail feathers, while female's feathers extend from the lower back to tail and those on tail ends are pale white.
Distribution: It is mainly distributed in Sichuan, northwestern Yunnan, southeastern Xizhang, southeastern Gansu, and southern Qinghai.
Protection class: Vulnerable species on the IUCN Red List, and the wildlife under state first-class key protection in China.

绿尾虹雉

英文名：Chinese monal
拉丁学名：*Lophophorus lhuysii*
分类：鸡形目 - 雉科 - 虹雉属
寿命：不详
食物：主要以植物的嫩芽、嫩叶、果实、种子、根茎为食。
外形特点：雄鸟具有紫色羽冠，下背白色，尾羽蓝绿色；雌鸟下背到尾上覆羽，尾端浅白色。
分布：主要分布于四川、云南西北部、西藏东南部、甘肃东南部和青海南部一带。
保护等级：IUCN 红色名录易危物种，中国国家一级重点保护野生动物。

我全身闪烁着金属光泽。
I am glowing with a metallic sheen.

藏酋猴

英文名： Tibetan macaque
拉丁学名： *Macaca thibetana*
分类： 灵长目 – 猴科 – 猕猴属
寿命： 20 岁左右
食物： 杂食性，但以植物为主，以多种植物的叶、芽、果、枝及竹笋为食，兼食昆虫、蛙、鸟卵等动物性食物，有时到农作物区取食。
外形特点： 身体粗壮，尾巴短，毛发长而浓密，背部深褐色，腹部颜色较浅。成年雄猴的面部为肉色，眼周为白色；雌猴的面部带有红色，眼周为粉红色。
分布： 中国特有种，广泛分布于中国中部和东南部。
保护等级： IUCN 红色名录近危物种，中国国家二级重点保护野生动物。

Tibetan macaque

English name: Tibetan macaque
Scientific name: *Macaca thibetana*
Classification: Primates - Cercopithecidae - *Macaca*
Lifespan: Around 20 years old
Food: It is an omnivorous animal, mainly feeding on leaves, buds, fruits, and branches of a variety of plants as well as and bamboo shoots, and also feeds on animal foods such as insects, frogs, and bird eggs, and sometimes feeds on crops.
Appearance: It has a stout body, short tail, and long and dense fur. Its back is dark brown white its belly is light in color. Adult males have flesh-colored faces with white eye contours, while females have red faces with pink eye contours.
Distribution: It is endemic to China and widely distributed in central and southeastern China.
Protection class: Near Threatened species on the IUCN Red List, and the wildlife under state second-class key protection in China.

> 我身材粗壮，尾巴短，是中国猕猴属动物中体形最大的一种。
> With a stout body and short tail, I am the largest of the Chinese macaques.

Golden pheasant

English name: Golden pheasant
Scientific name: *Chrysolophus pictus*
Classification: Galliformes - Phasianidae - *Chrysolophus*
Lifespan: 15 ~ 20 years old
Food: It eats flowers, fruits, leaves, buds, and seeds of various plants, as well as small insects.
Appearance: The male bird has red, orange, yellow, green, cyan, blue, and purple feathers.
Distribution: It is an endemic species of birds in China, widely distributed in the mountainous areas of central and western China.
Protection class: Least Concern species on the IUCN Red List, and the wildlife under state second-class key protection in China.

红腹锦鸡

英文名：Golden pheasant
拉丁学名：*Chrysolophus pictus*
分类：鸡形目 - 雉科 - 锦鸡属
寿命：15 ~ 20 岁
食物：各种植物的花、果、叶、芽、种子，以及小型昆虫等。
外形特点：雄鸟羽色赤橙黄绿青蓝紫俱全，色彩华丽。
分布：中国鸟类特有种，广泛分布于中国中西部山地。
保护等级：IUCN 红色名录无危物种，中国国家二级重点保护野生动物。

我是雄性，我的羽色赤橙黄绿青蓝紫俱全，色彩华丽。

I am a male golden pheasant with feathers in red, orange, yellow, green, cyan, blue, and purple.

大熊猫的同域伴生植物
Sympatric Companion Plants of Giant Pandas

珙桐

拉丁学名：*Davidia involucrata*
分类：蓝果树科 – 珙桐属
生长海拔：1500~2200米
花期：4月
果期：10月
保护等级：中国特有种，中国国家一级重点保护植物。

植物活化石

生长在深山云雾中的珙桐是古老的孑遗植物，有"植物活化石"之称，是国家8种一级重点保护植物之一，对研究古植物区系和系统发育具有重要的科学价值。珙桐曾在世界上广泛分布，经过第四纪冰期后，现在的珙桐只在中国南方零星分布，尤以四川为多。

人们叫我"鸽子树"

珙桐花开时，两片白色苞叶俯垂，酷似展翅飞翔的白鸽，因此被称为"鸽子树"。1869年，将大熊猫介绍给西方的法国传教士阿尔芒·戴维将在四川采到珙桐标本寄往巴黎。珙桐属的属名 *Davidia* 就来自他的名字。1904年珙桐被引入欧洲和北美洲，成为世界闻名的观赏树。

Davidia involucrata

Scientific name: *Davidia involucrata*
Classification: Nyssaceae - *Davidia*
Growth altitude: 1,500 ~ 2,200 meters
Flowering phase: April
Fruiting phase: October
Protection class: A plant under state first-class key protection endemic to China.

Living fossil

As one of the 8 plants under state first - class key protection in China, the *Davidia involucrata*, which grows in mountains covered by clouds and mists, is a species of relic plant and recognized as a "living fossil", with important scientific value for the study of paleoflora and phylogeny. *Davidia involucrata* were distributed widely in the world. However, after the quaternary glaciation, this species can only be seen in southern China, especially in Sichuan.

Dove tree

When the tree blossoms, the two white bracts hang down, and they resemble a white dove with wings outspread. Therefore, it is called "dove tree". In 1869, Armand David, the French missionary who introduced giant pandas to the western world, sent the specimen of *Davidia involucrata* he collected in Sichuan to Paris. The genus name *Davidia* comes from his name. In 1904, it was introduced to Europe and North America and became a world-famous ornamental tree.

红豆杉：植物界的大熊猫

拉丁学名： *Taxus chinensis*
分类： 柏目 – 红豆杉科 – 红豆杉属
生长海拔： 1000~1200 米以上
花期： 2~3 月
保护等级： 中国特有种，IUCN 红色名录易危物种，中国国家一级重点保护植物。

保护价值

红豆杉是第四纪冰川时期遗留下来的古老树种，因生长缓慢而极为稀少，被誉为"植物界的大熊猫"。红豆杉的药用价值在 2000 多年前就得到了充分肯定，成书于明朝的古医典《本草纲目》对红豆杉作过详细记载。现在，用红豆杉提取物紫杉醇制成的抗癌药物在医疗上得到了广泛应用。

Taxus chinensis: "Giant Pandas" of the Plant World

Scientific name: *Taxus chinensis*
Classification: Cupressales - Taxaceae - *Taxus*
Growth altitude: 1,000 ~ 1,200 meters and higher
Flowering phase: February and March
Protection class: Vulnerable species on the IUCN Red List, and a plant under state first-class key protection endemic to China.

Protection value

The *Taxus chinensis* is an ancient tree species left over from the quaternary glaciation, which is extremely rare due to its slow growth, and is known as the "giant panda of the plant world". The medicinal value of *Taxus chinensis* has been affirmed more than 2,000 years ago, and recorded in the ancient medical book *Compendium of Materia Medica*, written in the Ming Dynasty. Now, anti-cancer drugs made with paclitaxel, a *Taxus chinensis* extract, are widely used in medical treatment.

连香树：第三纪孑遗植物

拉丁学名： *Cercidiphyllum japonicum*
分类： 虎耳草目 – 连香树科 – 连香树属
生长海拔： 650~2700 米
花期： 4 月
果期： 8 月
保护等级： IUCN 红色名录无危物种，中国国家二级重点保护植物。

保护价值

连香树是第三纪古热带植物的孑遗种，是较古老原始的木本植物，现在已濒临灭绝。连香树在中国和日本间断分布，对于阐明第三纪植物区系起源以及中国与日本植物区系的关系有着较大的科研价值。

Cercidiphyllum japonicum: Relic Plant of the Tertiary Period

Scientific name: *Cercidiphyllum japonicum*
Classification: Saxifragales - Cercidiphyllaceae - *Cercidiphyllum*
Growth altitude: 650 ~ 2,700 meters
Flowering phase: April
Fruiting phase: August
Protection class: Least Concern species on the IUCN Red List, and a plant under state second-class key protection in China.

Protection value

The *Cercidiphyllum japonicum* is a relic species of tertiary paleotropical plants, and it is an ancient and primitive woody plant, but it is now on the verge of extinction. Its intermittent distribution in China and Japan has great scientific value for elucidating the origin of tertiary flora and the relationship between the flora of China and Japan.

Cephalotaxus oliveri: All are useful

Scientific name: *Cephalotaxus oliveri*
Classification: Cephalotaxales - Cephalotaxaceae - *Cephalotaxus*
Growth altitude: 300 ~ 1,800 meters
Flowering phase: March and April
Protection class: Vulnerable species on the IUCN Red List and a plant under state second-class key protection in China.

Protection value

The *Cephalotaxus oliveri* is a relic plant. It has special leaf shape and arrangement, which is obviously different from other plants of the same genus, and is of scientific significance for the study of paleoflora and the systematic classification of the *cephalotaxus*. At the same time, its leaves, branches, seeds, and roots can extract a variety of plant alkaloids, which have certain curative effects on the treatment of leukemia and lymphosarcoma.

篦子三尖杉：全身都是宝

拉丁学名：*Cephalotaxus oliveri*
分类：三尖杉目 – 三尖杉科 – 三尖杉属
生长海拔：300~1800 米
花期：3~4 月
保护等级：IUCN 红色名录易危物种，中国国家二级保护植物。

保护价值

篦子三尖杉是子遗植物，它的叶形及其排列极为特殊，与同属的其他植物有明显的区别，对于研究古植物区系和三尖杉属系统分类具有科学意义。同时，它的叶、枝、种子、根可提取多种植物碱，对治疗白血病及淋巴肉瘤等有一定疗效。

Tetracentron sinense: Ancient angiosperm

Scientific name: *Tetracentron sinense*
Classification: Trochodendrales - Trochodendraceae - *Tetracentron*
Growth altitude: 1,700 ~ 3,500 meters
Flowering phase: June ~ July
Fruiting phase: September ~ October
Protection class: A plant under state second-class key protection in China.

Protection value

Tetracentron sinense is a living fossil left over from the tertiary period. It is a primitive angiosperm with ancient origin, isolated system locations and a special ecological environment, which is of great value for the study of the origin of angiosperms.

水青树：古老的被子植物

拉丁学名：*Teracentron siense*
分类：昆栏数目 – 昆栏树科 – 水青树属
生长海拔：1700~3500 米
花期：6~7 月
果期：9~10 月
保护等级：中国国家二级重点保护植物。

保护价值

水青树是第三纪遗留下来的植物"活化石"。它是一种原始的被子植物，起源古老，系统位置孤立，生态环境特殊，对研究被子植物的起源具有重要的价值。

Chapter 7　第七章

Join the Future
创享未来

绿水青山，无骄永续。
践行环保，你我同行。

Meandering across the countless mountain peaks and valleys, separated by crystal clear mountain rivers and lakes, are pandas the pride of the forests, who will survive forever. Make efforts jointly to go green.

我们家族从 800 万年前走来，在适应大自然的过程中，创造了动物族群延续的奇迹。现在，我们已成为中国的一张名片，影响力遍及人们生活的各个方面。作为世界自然基金会的标志，我们用自己的"熊猫力量"唤起大众对物种保护的关注，让大家看到生态保护的必要性，看到生物多样性的重要性。

　　在广袤无限的大自然面前，一个人，一个组织，乃至一个国家的力量都显得无比渺小，但是每一个你都是不可或缺的生态护卫者。

　　谢谢大家关注我们！请大家试着多了解我们的真正需求，用科学的方式保护我们。珍惜和爱护地球资源，就是对我们最好的保护！

　　这个世界，有你们也有我们才够美好，你说对吗？！

>>

Over the 8 million year' course of evolution to better adapt to nature, we turned out to be species with a miraculously-long survival. Now, we have become one of China's symbols, whose influence has reached every aspect of their lives. As the logo of the WWF, we, a previously endangered species, intend to arouse public attention to species conservation by exposing them to the necessity of ecological conservation and the importance of biodiversity.

Compared with the vast and infinite nature, the power of a person, an organization, or even a country is dwarfed, but each of you is an indispensable guardian of the ecological balance.

Thank you for standing with us! I would appreciate it if you could try to understand more about our real needs and conserve us with scientific approaches. Cherishing the earth's resources proves to be the best conservation for us!

This world will be better with both you and us. Don't you think so?

嗨，我是小川。
Hi, I'm Xiao Chuan.

>>

Chengdu
Giant Panda
Museum
Exhibition Hall 7
Join the Future

成都大熊猫博物馆 第七展厅 创享未来

Influence of Giant Pandas
大熊猫的影响力

Nowadays, giant pandas are beloved by people from different countries, and their fans can be found all over the world. Elements of giant pandas can be seen in cultural exchanges, sports events, film and television, music, daily life, and other aspects. The giant panda is a superstar of the animal kingdom.

如今，大熊猫受到了世界各国人民的喜爱，"粉丝"遍布全球。在文化交流、体育赛事、影视音乐、日常生活等各个方面，我们都能看到与大熊猫相关的元素。大熊猫不愧是动物界的"顶级流量"。

大熊猫与文化艺术
Giant Pandas and Culture and Arts

As the national treasure of China, giant pandas are of great cultural and artistic values. Cultural and artistic works with the theme of giant pandas are blooming at home and abroad, showing a trend of diversified development.

作为中国的"国宝",大熊猫蕴藏着巨大的文化和艺术价值。现在,国内外以大熊猫为主题的文化艺术作品百花齐放,呈现出多元化发展的态势。

歌曲《熊猫咪咪》

《熊猫咪咪》发行于1984年,是一首为拯救"国宝"大熊猫而创作的公益歌曲。20世纪80年代初,岷山山系和邛崃山系的箭竹大面积开花枯死,大熊猫面临粮食危机。为号召人们一起保护大熊猫,艺术家创作了这首歌曲。

Song - *Giant Panda Mi Mi*

It is a public welfare song released in 1984 about saving the "national treasure" giant panda. In the early 1980s, arrow bamboos in the Min and Qionglai Mountains largely bloomed and withered, and giant pandas were facing a food crisis. The artist created this song as a call for people to work together to protect the giant pandas.

音乐专辑《超口爱行大运》

全球第一张以大熊猫为主角的音乐专辑。

Music Album - *Super Q*

It is the first music album in the world featuring giant pandas.

歌曲《伴你回家》

全球第一首以大熊猫野化放归研究为主题的单曲。随着圈养大熊猫种群的壮大,复壮野外种群,避免破碎化的栖息地中大熊猫小种群的灭绝被提上了日程。通过大熊猫野化放归研究,圈养大熊猫踏上了回归野外家园之路。

Song - *Accompanying You Home*

It is the world's first single about giant panda rewilding research. As the captive giant panda population grows, restoring the wild population and avoiding the extinction of small giant panda populations in fragmented habitats is on the agenda. By studying the rewilding of giant pandas, captive giant pandas have embarked on the road to return to their homes in the wild.

电影《熊猫历险记》

这是一部 1983 年上映的以保护大熊猫为题材的儿童电影,讲述了白马藏医老爷爷和他的孙子小凤救助大熊猫平平的感人故事。

Film - *Adventure of a Panda*

It is a children's film about the conservation of giant pandas released in 1983. It narrates a touching story of a Baima Tibetan doctor and his grandson Xiao Feng who rescued the giant panda Ping Ping.

纪录片《大熊猫 51 的故事》

这是 2012 年在日本上映的一部纪录片,讲述了 2006 年在成都大熊猫繁育研究基地出生的超级早产儿"51"的故事。它出生时体重仅有 51 克,仅为正常大熊猫初生幼仔体重的三分之一,是当时世界上出生时体重最轻的大熊猫。纪录片以"51"的成长过程为主线,全面记录了大熊猫繁育、成长的过程,突出表现了大熊猫母子和兄弟姐妹间感人至深的亲情。

Documentary - *The Story of Giant Panda 51*

It is a documentary released in Japan in 2012. It tells the story of the cub 51, which was born very premature in 2006 at the Chengdu Research Base of Giant Panda Breeding. It weighed only 51 grams at birth, only one-third the weight of a normal giant panda newborn cub, making it the world's lightest giant panda at birth at the time. The documentary narrates 51's growth, comprehensively recording the process of breeding and raising giant pandas, and highlights the touching family affection between giant panda mothers, children, and siblings.

纪录片《熊猫 熊猫》

3D 纪录电影《熊猫 熊猫》采用动画与实景拍摄结合的方式,追溯大熊猫族群 800 万年以来的演化历程,解读大熊猫与生俱来的生存智慧。

Documentary - *Panda Panda*

The 3D documentary film *Panda Panda* uses a combination of animation and live-action shooting to trace the evolution of the giant panda population over 8 million years and interpret the innate survival wisdom of giant pandas.

音乐剧《熊猫童话》

这是成都大熊猫繁育研究基地以中国—丹麦大熊猫文化交流活动为契机推出的国内首部大熊猫主题全英文原创儿童音乐剧。该剧以大熊猫"星二""毛二"赴丹麦开展大熊猫科研保护国际合作为背景，在创作中融合了中丹两国丰富的文化元素，由两国少儿演员共同演出。

Musical theater - *Panda Fairy Tale*

This is the first domestic original English children's musical theater about giant pandas launched by the Chengdu Research Base of Giant Panda Breeding during the China-Denmark Giant Panda Cultural Exchanges. Inspired by the giant pandas Xing Er and Mao Er going to Denmark for international cooperation on giant panda scientific research and conservation, the play integrates both Chinese and Danish cultures, and was acted by both children actors from China and Denmark.

话剧《幻镜1·冰川灵兽》

这是全球首部大熊猫历史话剧。该剧以当代大熊猫饲养员回望大熊猫进化史为切入点，讲述了在30万年前中更新世时期的大背景下，大熊猫、剑齿虎、猛犸象与早期猿人应对自然环境的大灾变、努力生存的历史。

Drama - *Fantasy 1 · Spirit Animal in the Ice Age*

This is the world's first historical drama about giant pandas. The drama takes contemporary giant panda breeders as the starting point to look back at the evolutionary history of giant pandas, and tells the story of the history of giant pandas, saber toothed tigers, mammoths, and early hominids responding to natural disasters and striving for survival in the middle Pleistocene period 300000 years ago.

杂志《看熊猫》

《看熊猫》杂志2004年创刊，是全球唯一一本公开发行的、以大熊猫命名的文化科普类杂志。杂志内容涵盖熊猫的故事、熊猫与人类的故事、熊猫与自然的故事、熊猫与世界的故事等主题，集知识性、趣味性、互动性于一体，被誉为"动物版的国家地理"。

Magazine - *Giant Panda*

Founded in 2004, the magazine *Giant Panda* is the only publicly-released cultural science magazine in the world named after the giant panda. It covers stories of giant pandas, giant pandas and mankind, giant pandas and nature, and giant pandas and the world. Reputed as the animal's own National Geographic, it integrates knowledge, fun, and interaction.

大熊猫与体育
Giant Pandas and Sports

盼盼

1990 年，北京亚运会吉祥物

1990 年在北京举办的第 11 届亚运会是中国第一次举办综合性国际体育大赛。北京亚运会以大熊猫盼盼为吉祥物，"盼盼"寓意"盼望和平、友谊，盼望迎来优异成绩"。它是以福州熊猫世界的大熊猫"巴斯"为原型设计的。

Pan Pan

Mascot for the 1990 Beijing Asian Games

The 11th Asian Games in Beijing in 1990 was the first comprehensive international sport competition held by China. The Beijing Asian Games chose the giant panda Pan Pan as its mascot. The mascot was designed based on Ba Si, the giant panda at Fuzhou Panda World, and carried the meaning of "looking forward to peace, friendship, and excellence".

大熊猫"巴斯"
Giant panda Ba Si

晶晶

2008 年，北京夏季奥运会吉祥物

福娃晶晶是 2008 年在北京举办的第 29 届夏季奥运会的吉祥物之一，其原型是成都大熊猫繁育研究基地的大熊猫"毛毛"。晶晶象征着人与自然的和谐共存，向世界各地的人们传递了友谊、和平、积极进取的精神，以及人与自然和谐相处的美好愿望。

Jing Jing

Mascot for the 2008 Beijing Summer Olympics

Jing Jing is one of the five mascots for the 29th Summer Olympics held in Beijing in 2008. Jing Jing's design was based on Mao Mao, a giant panda at the Chengdu Research Base of Giant Panda Breeding, and symbolizes the harmonious co-existence of mankind and nature. This mascot conveys friendship, peace, a positive spirit, and desire that mankind live in harmony with nature.

大熊猫"毛毛"
Giant panda Mao Mao

冰墩墩

2022 年，北京冬季奥运会吉祥物

冰墩墩是 2022 年在北京举办的第 24 届冬季奥运会的吉祥物之一。冰墩墩将大熊猫形象与富有超能量的冰晶外壳相结合，头部外壳造型的灵感来自冰雪运动头盔，装饰着彩色光环，象征冬奥会运动员强壮的身体、坚忍不拔的意志和鼓舞人心的奥林匹克精神。

Bing Dwen Dwen

Mascot for the 2022 Beijing Winter Olympics

Bing Dwen Dwen is one of the two mascots for the 24th Winter Olympics held in Beijing in 2022. Bing Dwen Dwen is a giant panda wearing an ice shell. The ice shell was inspired by a helmet in winter sports and the colorful halo around its face embodies the strength and willpower of athletes as well as the inspiring Olympic spirit.

蓉宝

2023 年，成都世界大学生运动会吉祥物

蓉宝是 2023 年在成都举办的第 31 届世界大学生运动会的吉祥物，它的原型是成都大熊猫繁育研究基地的大熊猫"芝麻"。蓉宝将憨态可掬的大熊猫形象与热情的火焰元素融为一体，全方位多视角地凸显了"火"这一天府文化的重要标签。它手持大运火炬，以欢快奔跑的姿态向全世界传递着青春和活力。

Rongbao

Mascot for the 31st FISU World University Games in Chengdu in 2023

Rongbao is the mascot for the 31st FISU World University Games held in Chengdu in 2023. It was based on Zhi Ma, a giant panda at the Chengdu Research Base of Giant Panda Breeding. Rongbao integrates the adorable image of giant panda with the passionate element of flame to highlight the important Tianfu Culture element fire from all angles and perspectives. Rongbao holds a torch in its hand and conveys youth and vitality to the world with a cheerful attitude in a runner's pose.

大熊猫"芝麻"
Giant panda Zhi Ma

大熊猫与日常生活
>> "Giant Pandas" in Our Daily Life

第七章／创享未来

大熊猫元素早已走进我们的日常生活，从文创产品，到生活用品、城市地标，随处可见"大熊猫"的身影。

Giant panda elements have entered our daily lives for a long while. "Giant pandas" can be seen everywhere, from cultural and creative products, to daily necessities and city landmarks.

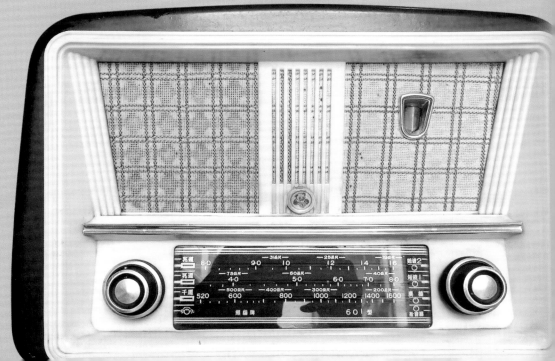

Protecting the environment is to conserve Giant Pandas

保护环境就是保护大熊猫

为保护以大熊猫为代表的生物多样性，科研人员在大熊猫繁育、建立野生大熊猫遗传种质资源库等方面全力攻关，"奶爸""奶妈"们每天悉心照顾大熊猫的日常起居。那么，普通人能不能为保护大熊猫做点儿力所能及的事呢？答案是肯定的。不积跬步，无以至千里；不积小流，无以成江海。在日常生活中践行各种环保行为，选择可持续的生活方式，这就是在为共建"地球生命共同体"贡献力量。

In order to protect the biodiversity represented by giant pandas, researchers have not made full efforts in panda breeding, establishing a genetic resource bank for wild giant pandas, and other areas. The keepers take good care of the daily lives of giant pandas every day. Can every ordinary person do something within their ability to protect giant pandas? The answer is yes. Without accumulating small streams, one cannot form a river or sea. Without accumulating small steps, one cannot reach a thousand miles. Practicing various environmental protection behaviors in daily life and choosing a sustainable lifestyle is contributing to the construction of a "community of life on Earth".

这些行动很环保
>> Eco-friendly Behaviors

日常生活中我们应该怎么做？

What can we do in our daily life?

1. Plant and breed native species, and reject alien species;

2. Reduce the use of disposable goods and avoid over-packaged products;

3. Green travel and take public transportation as much as possible;

4. Use rechargeable batteries and recycle disposable batteries;

5. Stop actions that damage the environment;

6. Refrain from eating or buying wild animals and plants and their products;

7. Advocate for green life, save energy, and do not consume excessively.

1. 种植、养殖本地物种，拒绝外来物种；

2. 尽量减少使用一次性物品，避免购买过度包装的商品；

3. 绿色出行，尽量乘坐公共交通工具；

4. 使用充电电池，注意回收利用普通电池；

5. 主动制止破坏环境的行为；

6. 坚决不吃野生动植物，拒绝购买和使用野生动植物及其制品；

7. 倡导绿色生活，节约能源，不过度消费。

参观动物园时我们应该怎么做?

4. 不向动物投食;

5. 不敲打动物笼舍或玻璃;

6. 不带宠物入园;

7. 不参与动物商业合影。

1. 不追逐动物;

2. 不用奇怪的声音和动作惊吓动物;

3. 拍照时关闭闪光灯;

What should we do when visiting a zoo?

1. Do not chase the animals;

2. Do not scare animals with strange noises or behaviors;

3. Turn off the flash when taking photos;

4. Do not feed the animals;

5. Do not knock on animal cages or glass;

6. Do not bring your pets;

7. Do not take photos of animals for commercial purpose.

户外旅行时我们应该怎么做？

What should we do when enjoying the outdoors?

1. Do not travel in unmapped areas;

2. Do not litter, and take away any garbage;

3. Do not approach, touch, or catch any wild animals;

4. Do not destroy or take away anything, such as flowers, grass, stones, branches, or shells;

5. Do not start a fire in the wild or litter cigarette butts;

6. Do not yell at any animals you see or disturb their normal activities;

7. Do not bring any seeds, plants, or animals into the wild.

1. 不在未经规划的区域开展旅游活动；

2. 不乱扔垃圾或留下任何废弃物；

3. 不要试图靠近、触摸或抓走任何野生动物；

4. 不破坏或带走任何自然物，如花草、石头、树枝、贝壳等；

5. 不要在野外生火，也不要乱扔烟头；

6. 不向你看到的动物吼叫，不干扰野生动物的正常生活；

7. 不要把任何种子、植物或动物带到野外。

环保行动者们的心声
>> Voices of Environmental Activists

> 有了健康的环境，
> 我们才会有健康的身体。
> Only in a healthy environment
> can we have a healthy body.

隆芯一号

保护宣言：美丽中国，我是行动者！

我是一名生态保护从业者，长期坚守在环保公益领域，以垃圾分类为切入点带动公众参与各项环保行动。多年如一日，我坚持让垃圾分类的意识深入人心。为了美丽的世界，我会坚持下去。

Longxin Yihao

Slogan: Act to build a beautiful China!

As an ecological protector, I have been committed to environmental public welfare and led the public in participating in various environmental protection actions that began with garbage classification. For many years, I have insisted on bringing awareness to garbage classification. For the sake of our beautiful world, I will continue.

曦文

保护宣言：保护环境，敬畏自然，坚持可持续发展，从我开始。以生命之名，为自然发声！

我是一名科普志愿者，从学生时代起，开始参加各类环保活动，日常生活中也会通过各种行动来践行环保。同时，我还会在科普讲座中分享自己的环保体悟，并且呼吁身边的同学加入环保行动。

Xiwen

Slogan: Start with me to protect the environment and respect nature and adhere to sustainable development. Speak for nature in the name of life!

I am a science volunteer. I have been participating in various environmental protection activities since I was a student, and I also put environmental protection into practice through various actions in my life. Meanwhile, I also share my environmental protection experiences and understandings in popular science lectures, and call on my classmates to take environmental protection actions.

带鱼

保护宣言：自然从未被征服，他只是给予了我们包容。守护自然，守护家园！

我是一名营地教育工作者，工作之余也会尽最大努力向公众普及野生动物保护等科普知识。2006年4月在四川省冕宁县冶勒乡，我通过法律普及和现场教育，成功从猎人手中救下了国家二级保护动物小熊猫。守护野生动物和它们的家园，是我的责任！

Hairtail

Slogan: Nature has and will not be conquered. It merely tolerates us. Protecting nature is protecting our homeland!

I am a camp educator, and I do my best to educate the public about wildlife conservation and other scientific knowledge outside of work. In April 2006, I, through legal popularization and on-site education, rescued a red panda, an animal under state second-class protection in China, from hunters in Yele Township, Mianning County, Sichuan Province. It is our responsibility to protect wild animals and their homeland!

江连长

保护宣言：大自然可以不需要人类，可是人类需要大自然！

我是一名自然教育从业者，2013年初进入这个行业。我的教育理念是"让孩子像孩子那样去生长，让我们最终都活成孩子，让一切都自然而然"。我喜欢向孩子和社会传递动物保护和环境保护的美好理念。

Company Commander Jiang

Slogan: Nature can be great without mankind, but mankind cannot live without nature!

I have become a nature educator in early 2013. My educational philosophy is, "Let children grow like children; let us all live like children to the end, and let everything happen naturally". I love to convey the beautiful idea of animal conservation and environmental protection to children and society.

魏淇澳

保护宣言：用手中的画笔保护正在消失的自然。

我是一名大学生，从小便喜爱自然观察和艺术。作为环境科学的学子，我在大学里接触了许多野外实践的机会，对保护自然有了更专业的了解，并且我也发现我的画笔能够吸引更多的人一起来关注自然，保护自然。

Wei Qi'ao

Slogan: Protect the disappearing nature with my brush.

I am a college student, who has loved nature observation and art since childhood. As a student of Environmental Science, I have practiced in the wild many times and have acquired a more professional understanding of nature conservation. I have found my painting can move more people to care about and protect nature.

成都麻雀

保护宣言：有了健康的环境，我们才会有健康的身体。

我是一名服务业从业者。我从小喜欢看《动物世界》《人与自然》《国家地理》等电视节目，久而久之产生了保护野生动物的想法，在生活中也会通过各种行动来践行环保。我有幸成为成都观鸟会活动的志愿者，为公众讲解鸟类知识，也为保护自然奉献出了自己的一分力量。

Chengdu Sparrow

Slogan: Only in a healthy environment can we have a healthy body.

I am a service industry practitioner. I love to watch *Animal World*, *Human and Nature*, *National Geographic*, and other TV programs since I was young. Over time, I have developed the idea of protecting wild animals, and I protect the environment through various actions in my life. I was fortunate to be a volunteer at the Chengdu Bird Watching Society to explain knowledge about birds to the public, contributing to the protection of nature.

人类保护自然，
就是在保护人类自己。

When human beings protect nature, they protect themselves.

大熊猫守护先锋

保护宣言：不是自然为你奉献了什么，而是你为自然付出了多少！不忘绿色初心，牢记保护使命！

我是一名大熊猫自然保护区的巡护员，参加大熊猫保护工作已有19年。在远离城市的保护区工作虽然很艰苦，但是我仍然对事业有着无限赤诚。我一直守护在这里，为的是把上天给我们的馈赠在这一代保护人手里继续传承。

Giant Panda Guardian Pioneer

Slogan: Ask not what nature can do for you; ask what you can do for nature! Maintain a green planet and practice conservation!

I am a ranger in the giant panda nature reserve and have engaged in giant panda conservation for 19 years. Working in a reserve far from the city is hard, but I am still passionate about what I do. I have been guarding this place in order to pass on the gift of nature to future generations.

强风吹拂

保护宣言：低碳环保，还地球一片绿色，还天空一片蔚蓝。

我是一名公交车驾驶员。2018年，我开始驾驶新能源公交车。新能源公交车有漂亮的车身，舒适的车内环境，行驶时没有发动机的轰鸣声，零排放，低碳，节能，环保，安全，便捷，太棒了！我为自己能从事"绿色驾驶"感到骄傲。

Wind-blowing

Slogan: Low carbon and environment protection make the planet green and the sky blue again.

I am a bus driver. Since 2018, I have driven the new energy bus with a beautiful bodywork and comfortable environment. When the bus is running, there is no roar of the engine and no emissions. It conserves energy, is environmentally friendly, safe and convenient. It is awesome! I am so proud to be engaging in "green driving".

余芳

保护宣言：保护环境，保洁自然，就是保护自己的家园。

我是一名普通的环卫工人。早起晚睡，为了城市的清洁每天穿梭于大街小巷。我的工作包含清扫道路、捡拾烟头垃圾、制止和劝导破坏环境卫生的行为等。美好的城市环境不止需要我们环卫工人，还需要大家共同努力、共同建设、共同维护。

Yu Fang

Slogan: Protecting the environment and nature protects our homeland.

I am a sanitation worker. I clean the city streets from dawn until dusk, sweeping road, picking up cigarette butts and preventing and discouraging behavior that hurts the environment. A beautiful urban environment not only needs our sanitation workers, but also needs everyone's joint efforts, joint construction, and joint maintenance.

Jeroen Jacobs

保护宣言：分享您与全球各地大熊猫的故事，一同拯救地球。

1987年的夏天，两只大熊猫来到我的家乡比利时安特卫普。我十分痴迷于这对来自中国的大熊猫，于是，在2000年10月15日，我建立了一个大熊猫网站，与人们分享关于大熊猫的信息和新闻，同时向世界传递大熊猫保护的知识和大熊猫文化。

Jeroen Jacobs

Slogan: Share your stories with giant pandas and let's protect the earth together.

In the summer of 1987, two giant pandas came to my hometown, Antwerp, Belgium, and I have been obsessed with them ever since. I established a giant panda website on October 15, 2000 to share information and news about giant pandas and spread information about giant panda conservation and culture.

马鸿瀚

保护宣言：鸟儿，鱼儿，树儿，花儿，都是生态系统的一分子。保护了它们，也就保护了我们的家园。

我家几代居住在大熊猫自然保护区旁边的村子里，随着保护区的建立，我们从靠山吃山变成了环保事业的参与者。我积极参加保护区组织的社区活动，学习可持续的生产生活方式，向周边村民宣传关于珍稀动植物的知识和保护政策，帮助村民们提高了主动保护珍稀动植物的意识。

Ma Honghan

Slogan: Birds, fish, trees, and flowers all are a part of the ecosystem. Protecting them is protecting our homeland.

My family has resided in the village next to the giant panda nature reserve for generations. With the establishment of the reserve, we have changed from relying on the mountains to participating in environmental protection. I have participated in community activities organized by the reserve, learned about sustainable production and lifestyle, disseminated knowledge and conservation policies about rare animals and plants to the surrounding villagers, and helped them become more aware of the initiative to protect rare animals and plants.

Even

保护宣言：大自然本不需要人类的保护。人类保护自然，就是在保护人类自己。

我是一名口腔医学专业教育工作者，同时也是一个环境保护的践行者。每当我看到破坏环境的行为时，我都会意识到我们的环保教育做得还不够。我愿意将环保作为我的终生使命，一直践行下去。希望大家都加入到环保行动中来，为共建美好家园出一份力！

Even

Slogan: Nature does not need our protection. When human beings protect nature, they protect themselves.

I am an educator of dentistry and a practitioner of environmental protection. Every time I see something that destroys the environment, I firmly believe that we haven't done enough in environmental education. I am willing to make environmental protection as my lifelong mission and continue to work on it. I hope everyone will join me in environmental protection actions and contribute to building a better home!

熊猫世界之旅结束啦！
你一定想了解更多关于我们家族的知识，
想见到真正的我。

那么，
欢迎来到成都大熊猫繁育研究基地，
走进大熊猫博物馆。

我在这里等着你！

The trip to the world of
giant panda is over!
You are certainly eager to know more about our
family and to meet the real me.

Thanks for coming to the
Chengdu Research Base
of Giant Panda Breeding,
and the Chengdu Giant
Panda Museum.

I'll be here waiting for you!

到达
End

后记及致谢
>> Postscript and Acknowledgments

2007年7月，我硕士研究生毕业后就来到熊猫基地工作，成为科普教育团队的一员，初定岗位为博物馆专业技术人员。学习生态学专业的我从此跟大熊猫和博物馆结下了不解之缘。

当时，新的大熊猫博物馆的主体建筑刚刚竣工，我的任务就是负责推进博物馆的布展工作。因为各种原因，布展工作推进得并不顺利，反复多次启动又暂停，终于在2018年进入到实质性阶段。

对初次涉足博物馆布展工作的我来说，一切都要从头学起。我参考之前老馆留下来的资料，又找到众多专家向他们咨询、请教，最后一口气写完了布展大纲，并编写出了布展技术要求。但是，招标过程中又遭遇了很多难题，最后经过三次努力，才在2019年8月确定了布展单位。整个布展项目有时任熊猫基地主任张志和研究员的指导，有佘轶副主任的帮助，在布展施工上有王革生副主任、廖骏部长和傅琰华科长的严格把关。我和团队一起撰写布展内容，与布展单位共同沟通布展方式，疫情三年，我们克服了重重障碍，终于在没日没夜的加班中于2020年11月完成了大熊猫博物馆的布展工作。经过几个月的内部调试和试运行，大熊猫博物馆终于在2021年3月3日正式对公众开放。

当天，包括中央电视台、新华社新媒体、四川省人民政府网等在内的15家媒体平台对大熊猫博物馆开馆进行了宣传报道，央视网熊猫频道（iPanda）、咪咕网、CGTN英语环球等直播平台的相关报道收获了4000万以上人次的关注。博物馆开放后，馆内的互动装置、多媒体展示、科普活动等激发了游客们的浓厚兴趣，受到社会各界的一致好评……我们团队的辛苦付出收获了丰硕的果实。

今天你来到成都大熊猫博物馆，进入大厅，第一眼就能看到一本巨型的多媒体装置书，这本书的名字就叫《熊猫世界》。博物馆开馆第一天，我就想，将来一定要编写一本同样叫《熊猫世界》的书，将博物馆里关于大熊猫的展陈内容广泛传播。2023年疫情结束后，工作日益繁忙，我只能利用下班后和周末的时间废寝忘食地撰写书稿，见缝插针地和编辑、设计师沟通出版方案，其间艰辛，一言不足以道之。经过大半年的努力，今天，大熊猫博物馆终于被"搬"到了纸上！我给这本书的定位是大熊猫的博物

I have been working at the Panda Base since graduating with a master's degree in July 2007 and have joined the science popularization team primarily working as a professional technician at the Chengdu Giant Panda Museum. As an ecology-major student, I have been tightly tied to giant pandas and the museum.

Back then, the main building of the new Giant Panda Museum had just been completed, and I was responsible for exhibition arrangements. For various reasons, the exhibition arrangements didn't go smoothly. After repeated starts and spurts, it finally entered the substantive phase in 2018.

I was new to museum exhibition arrangements and had to learn everything from the ground up. I referred to the materials in the old museum and sought advice from many experts. Then I finished the exhibition outline and listed the technological requirements for the exhibition. However, there were many difficulties during the bidding process. Finally, after three attempts, the exhibition organizer was decided in August 2019. While arranging the exhibition, researcher Zhang Zhihe, the then-director of the Panda Base, She Yi, deputy director of the Panda Base provided assistance. In the process of construction, Wang Gesheng, deputy director, Liao Jun, department head, and Fu Yanhua, section chief, monitored and reviewed the construction. My team and I worked together to write the content of the exhibition and communicate with the exhibition organizer about the arrangements. During the 3 years of the pandemic, we worked day and night and overcame many obstacles, and finally completed the installation of the Giant Panda Museum in November 2020. Eventually, it was officially open to the public on March 3, 2021 after several months of internal adjustments and trial operation.

On March 3, 2021, a total of 15 media platforms, including CCTV, Xinhua News Agency, and the website of the Sichuan Provincial People's Government, reported on the opening of the Chengdu Giant Panda Museum. Related news in iPanda of the CCTV Network, Migu Web, CGTN NEWS Plus, and other live-broadcasting platforms were viewed by more than 40 million people. After the opening, the interactive installations, multimedia displays, and popular science activities in the

书，这本书的每个章节都对应了大熊猫博物馆的一个展厅，但又在展陈内容的基础上进行了深度拓展，是介绍大熊猫比较全面的一本科普图书。

这本书的出版离不开各位领导、同事、专家老师，以及朋友们的支持。感谢四川省科学技术厅，没有他们提供的资金支持，这样一本以中英双语形式编排的大部头科普书要想出版绝非易事；感谢熊猫基地尹志东主任对图书出版的大力支持，从图书内容策划到后期的配套宣传规划，都有他的帮助与指导，特别感谢他为本书作序推荐；感谢我们科普团队的每一位小伙伴，从图书策划之初，大家就非常支持我的想法，积极协助我整理撰写相关资料；感谢熊猫基地侯蓉研究员、兰景超研究员、黄祥明研究员、吴孔菊高级畜牧师、李明喜正高级畜牧师、刘玉良研究员、齐敦武研究员、罗娌副研究员、沈富军研究员对大熊猫知识内容的把关，他们是真正的大熊猫专家；感谢熊猫基地聂溟飞部长、杨奎兴科长、许萍科长、魏玲科长、李洁科长、张玉均副科长对大熊猫宣传文化、国际交往方面内容的把关；感谢我的同事吴樱、James Edward Ayala 对书稿英文部分的专业校对；感谢我的同事毕温磊、苏小艳在资料与数据方面提供的支持与帮助；感谢我的好友西南民族大学文新学院秦丽副教授，她对书稿的润色，让我对本书顺利出版充满信心；感谢成都理工大学沉积地质研究院杨文光教授对本书地质相关内容的编写给予的指导与帮助；感谢巫嘉伟老师对书中大熊猫伴生动植物相关内容的编写给予的指导与

Museum have aroused a strong interest in visitors. The Museum has been well received by all walks of life... Our pain has turned to gains.

Once visitors enter the hall, they will see a giant multimedia book, named *The World of Panda*. On opening day, I thought that I would write a book named that in the future, which would enrich and spread the exhibition on giant pandas far and wide. Ever since the pandemic ended in 2023, I became increasingly busy. I could only spend my off-duty hours and weekends writing the manuscript and communicating with editors and designers about publishing plans. The hardships are beyond description. After more than half a year of hard work, the Giant Panda Museum has finally been "moved" to paper! I define this book as a science book of giant pandas. Each chapter corresponds to an exhibition hall at the Giant Panda Museum and enriches the content of the exhibition. It is a comprehensive popular science book that introduces the giant panda.

The book could never have been published without the support of leaders, colleagues, expert teachers, and other friends. First of all, I want to thank the Science & Technology Department of Sichuan Province. Without its funding, it would have been no easy task to publish such a large-scale science book in Chinese and English. Second of all, I want to thank Yin Zhidong, the director of the Panda Base, for supporting the publication, providing assistance from content planning to promotion. I also want to thank him for writing the preface. I want to thank every one

成都大熊猫博物馆鸟瞰
Bird's-eye view of Chengdu Giant Panda Museum

on our team for supporting my ideas and assisting me in organizing the materials at the beginning. I want to thank researchers Hou Rong, Lan Jingchao, and Huang Xiangming, vice-senior zootechnician Wu Kongju, senior zootechnician Li Mingxi, researchers Liu Yuliang, Qi Dunwu, and deputy researcher Luo Li and researcher Shen Fujun for checking the information as true giant panda experts. I also want to thank the department head Nie Mingfei, section chiefs Yang Kuixing, Xu Ping, Wei Ling, Li Jie, and deputy section chief Zhang Yujun for reviewing the content about giant panda culture and international exchanges. I want to thank my colleagues Wu Ying and James Edward Ayala for proofreading the English version and thank Bi Wenlei and Su Xiaoyan for providing related information and data. I want to thank my friend Qin Li, an associate professor at the School of Literature and Journalism of Southwest Minzu University for polishing the words and making me feel more confident; Yang Wenguang, a professor at the Institute of Sedimentary Geology of Chengdu University of Technology for advising on the geology content. I want to thank Wu Jiawei for advising on the content of companion animals and plants of the giant panda; Luo Yu for reviewing the content related to the China Conservation and Research Center for Giant Panda; China West Normal University for providing photos of professor Hu Jinchu; the China Conservation and Research Center for Giant Panda for providing the photos of giant pandas Jing Jing, Si Hai, and Ueno Zoo, Japan; the Strait (Fuzhou) Giant Panda Research and Exchange Center for providing the precious pictures of giant panda Ba Si. I want to thank director Xiu Yunfang and thank my colleague Cui Kai, a professional giant panda photographer, for providing many pictures he took days and years; Yong Yan'ge for providing the excellent pictures of wild giant pandas, which are difficult to take. I am deeply touched by his contributions. I want to thank Xue Xianliang for designing this book, who has never failed to pique my interest. As he says, if you want to do it, you have to make something interesting.

In the end, there is no possibility for me to focus on writing when not at work, on weekends and holidays without my family's support and patience. I hope children as old as my son will find this book interesting. I hope that you not only learn about giant pandas, but also spread giant panda culture, begin to protect biodiversity with small changes, and contribute to China's ecological civilization construction.

Jin Shuang
Chengdu, China
March, 2024

帮助；感谢罗瑜老师对书稿中涉及的中国大熊猫保护研究中心相关内容的审核；感谢西华师范大学提供的胡锦矗教授的相关照片；感谢中国大熊猫保护研究中心提供的大熊猫"京京"和"四海"以及日本上野动物园的照片；感谢海峡（福州）大熊猫研究交流中心提供的大熊猫巴斯的珍贵照片，特别感谢修云芳主任的支持；感谢我的同事崔凯对本书出版提供的大量摄影图片，他日复一日、年复一年对大熊猫进行拍摄，是专业的大熊猫摄影师；感谢雍严格老师慷慨提供的精彩的野外大熊猫照片，野外大熊猫难寻，照片更加难拍，雍老师为此付出的艰辛让人感动；感谢薛先良先生对本书的设计付出的辛苦，薛先生设计的作品总是很有趣，按照他的话来说要做就要做点儿有意思的东西出来。

最后，我能在下班后、周末、节假日心无旁骛地创作，离不开家人的支持与包容。希望这本书能受到与我儿子年纪相仿的青少年朋友的喜欢，我希望你们不仅了解大熊猫知识，更要努力去传播大熊猫文化，从点滴小事开始践行生物多样性保护行为，为中国的生态文明建设贡献自己的一分力量。

金　双
2024 年 3 月于成都

参考书目

[1]黄万波, 魏光飚. 大熊猫的起源[M]. 北京: 科学出版社, 2010.

[2]黄万波, 李华, 秦勇, 胡鑫, 廖克海. 大熊猫的前世今生[M]. 北京: 科学出版社, 2018.

[3]亨利·尼科尔斯. 来自中国的礼物——大熊猫与人类相遇的一百年[M]. 黄建强, 译. 北京: 生活.读书.新知三联书店, 2018.

[4]四川省地方志编纂委员会. 四川省志·大熊猫志[第八十四卷][M]. 北京: 方志出版社, 2018.

[5]国家林业和草原局. 全国第四次大熊猫调查报告[M]. 北京: 科学出版社, 2021.

[6]金双. 你不知道的大熊猫[M]. 成都: 四川少年儿童出版社, 2021.

[7]胡锦矗, 胡晓. 寻踪国宝——走近大熊猫家族[M]. 南京: 江苏科学技术出版社, 2013.

[8]胡锦矗. 山野拾零[M]. 北京: 科学出版社, 2020.

[9]张志和, 塞娜·贝可索. 大熊猫: 生·存[M]. 魏玲, 译. 重庆: 重庆大学出版社, 2014.

[10]魏辅文. 野生大熊猫科学探秘[M]. 北京: 科学出版社, 2018.

[11]张志和. 熊猫大百科[M]. 成都: 天地出版社, 2020.

[12]赵学敏. 大熊猫——人类共有的自然遗产[M]. 北京: 中国林业出版社, 2006.

[13]四川省地方志工作办公室, 四川省林业和草原局. 大熊猫图志[M]. 北京: 方志出版社, 2019.

[14]蔡鹏, 金双. 告白熊猫——大熊猫脱濒记[M]. 昆明: 云南科技出版社, 2021.

[15]张志和, 魏辅文. 大熊猫迁地保护理论与实践[M]. 北京: 科学出版社, 2006.

Bibliography

[1]Huang, Wanbo., Wei, Guangbiao. (2020). *The Origin of Giant Panda*. Science Press.

[2]Huang, Wanbo., Li, Hua., Qin, Yong., Hu, Xin., Liao, Kehai. (2018). *The Past and Present of Giant Panda*. Science Press.

[3]Nicholls, H. (2018). *The Way of the Panda: The Curious History of China's Political Animal*. SDX Joint Publishing Company.

[4]Sichuan Provincial Committee of Local Records Compilation. (2018). *Sichuan Province Records - Annals of Giant Panda [Volume 84]*. Local Records Publishing House.

[5]National Forestry and Grassland Administration. (2021). *The 4th National Survey Report on the Giant Panda in China*. Science Press.

[6]Jin, Shuang. (2021). *The Giant Panda You Don't Know*. Sichuang Children's Publishing house.

[7]Hu, Jinchu., Hu, Xiao. (2013). *Trailing the Giant Panda*. Phoenix Science Press.

[8]Hu, Jinchu. (2020). *Giant Pandas in the Wild*. Science Press.

[9]Zhang, Zhihe., Bexell, S. (2014). *Giant Pandas: Born-Survivors*. Publishing house of Chongqing Univesity.

[10]Wei, Fuwen. (2018). *Hope for the Giant Panda — Scientific Evidence and Conservation Practice*. Science Press.

[11]Zhang, Zhihe. (2020). *Encyclopedia of Pandas*. Tiandi Press.

[12]Zhao, Xuemin. (2006). *The Giant Panda — Natural Heritage of Humanity*. China Forestry Publishing House.

[13]Office of Sichuan Provincial Local Records., Sichuan Provincial Bureau of Forestry and Grassland. (2019). *Giant Panda Pictorial Annals*. Local Records Publishing House.

[14]Cai, Peng., Jin, Shuang. (2021). *Confession to Pandas — History of Giant Pandas Clearing Their Endangered Status*. Yunnan Science and Technology Press CO., LTD.

[15]Zhang, Zhihe., Wei, Fuwen. (2006). *Giant Panda Ex-situ Conservation Theory and Practice*. Science Press.